Troubleshooting HVAC-R Equipment

Jim Johnson

Jim - 489-1136

Delmar Publishers

I(T)P An International Thomson Publishing Company

Albany • Bonn • Boston • Cincinnati • Detroit • London • Madrid • Melbourne
Mexico City • New York • Pacific Grove • Paris • San Francisco • Singapore • Tokyo
Toronto • Washington

NOTICE TO THE READER

Publisher does not warrant or guarantee any of the products described herein or perform any independent analysis in connection with any of the product information contained herein. Publisher does not assume, and expressly disclaims, any obligation to obtain and include information other than that provided to it by the manufacturer.

The reader is expressly warned to consider and adopt all safety precautions that might be indicated by the activities herein and to avoid all potential hazards. By following the instructions contained herein, the reader willingly assumes all risks in connection with such instructions.

The publisher makes no representation or warranties of any kind, including but not limited to, the warranties of fitness for particular purpose or merchantability, nor are any such representations implied with respect to the material set forth herein, and the publisher takes no responsibility with respect to such material. The publisher shall not be liable for any special, consequential, or exemplary damages resulting, in whole or part, from the readers' use of, or reliance upon, this material.

Cover Image by Charles Cummings Advertising / Art Inc.

Delmar Staff
Publisher: Susan Simpfenderfer
Acquisitions Editor: Vernon Anthony
Developmental Editor: Jeanne Mesick
Project Editor: Patricia Konczeski
Production Coordinator: Karen Smith
Art/Design: Cheri Plasse

Printed in the United States of America.

For more information, contact:

Delmar Publishers
3 Columbia Circle, Box 15015
Albany, New York 12212-5015

International Thomson Publishing Europe
Berkshire House 168 - 173
High Holborn
London WC1V 7AA
England

Thomas Nelson Australia
102 Dodds Street
South Melbourne, 3205
Victoria, Australia

Nelson Canada
1120 Birchmount Road
Scarborough, Ontario
Canada M1K 5G4

International Thomson Editores
Campos Eliseos, Piso 7
Col Polanco
11560 Mexico D F Mexico

International Thomson Publishing GmbH
Königswinterer Strasse 418
53227 Bonn
Germany

International Thomson Publishing Asia
221 Henderson Road
#05 - 10 Henderson Building
Singapore 0315

International Thomson Publishing Japan
Hirakawacho Kyowa Building, 3F
2-2-1 Hirakawacho
Chiyoda-ku, Tokyo 102
Japan

8 9 10 11 xxx 10 09 08 07

Library of Congress Cataloging-in-Publication Data

Johnson, Jim, 1950 -
Troubleshooting HVAC-R equipment / by Jim Johnson.
p. cm.

ISBN-13: 978-0-8273-6392-2
ISBN-10: 0-8273-6392-3

1. Air conditioning—Equipment and supplies—Maintenance and repair. 2. Heating—Equipment and supplies—Maintenance and repair. 3. Ventilation—Equipment and supplies—Maintenance and repair. 4. Refrigeration and refrigerating machinery—Maintenance and repair. I. Title.
TH7687.7.J64 1996 95-35411
697'.0028'8—dc20 CIP

CONTENTS

SECTION THREE: TOOLS, EQUIPMENT, AND METERS

SECTION FOUR: HVAC-R EQUIPMENT

Tracking down a problem in an electrical or refrigeration system is a skill that must be developed on an individual basis after some basic concepts are understood, and in order to benefit most from the contents of this text, an effort on the part of the reader is necessary. While there are some things that can be learned in a passive mode, troubleshooting is not one of them.

In order to become an effective troubleshooter, a firm understanding of basic refrigeration and electrical concepts must be established before wiring diagrams can be interpreted or refrigeration system problems can be diagnosed. This is not to say that every time you begin to trace an electrical circuit you must stop to remember the theory of electromagnetism, but understanding this concept and eliminating the mystery behind it clears the way for a full understanding of how electrical energy is used to perform work in an HVAC-R system.

In the natural order of things, this one is for Shannon.

SECTION

1

REFRIGERATION FUNDAMENTALS

In order to effectively troubleshoot HVAC-R equipment, you must first become familiar with the basic principles of refrigeration. Understanding the laws of heat transfer and how they affect the operation of a refrigeration system allows you to proceed confidently when diagnosing and correcting system problems. Eliminating the "mystery" behind the refrigeration cycle is one of the building blocks in the foundation of good troubleshooting skills. Identifying the four basic components within a refrigeration system and knowing how they do their job is another.

Understanding and being able to apply this information can make the difference between a skilled technician who is able to effectively troubleshoot a system and a parts changer who relies on guesswork alone.

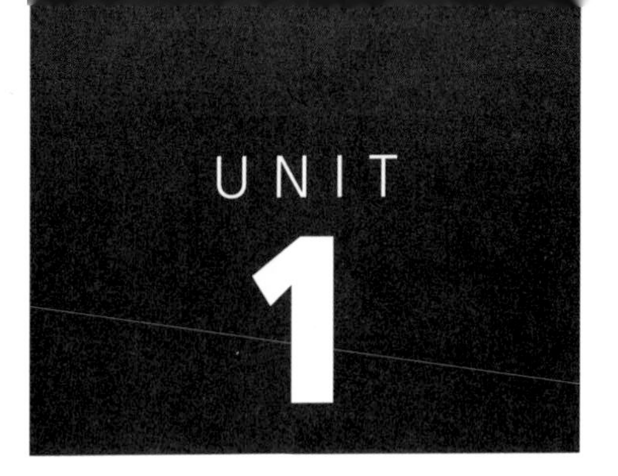

Heat Energy

Refrigeration is the transfer of heat from a place where it is not wanted to a place where it is unobjectionable. In other words, a refrigeration system does not put cold *into* an area, it takes the heat *out*. You are not kept comfortable in an air-conditioned movie theater because the cooling system is dumping cold air into the space, but rather because the air handling system has drawn the air through a coil in which a refrigerant has absorbed heat, then recirculated the cooler air to the conditioned space.

This process applies to any refrigeration system, be it a comfort cooling system, dairy case in a grocery store, walk-in cooler in a restaurant, or the refrigerator in your kitchen. To understand the process of refrigeration, you must first become familiar with some of the laws of thermodynamics. The term *thermodynamics* refers to heat transfer.

QUICK NOTE

The first law of thermodynamics is that heat always moves from a warmer surface to a cooler surface.

If you were to park a black automobile in Death Valley at high noon on the 20th of July, wait an hour, then place your palm on the hood of the car, you would experience a graphic illustration of the transfer of heat from a warmer surface to a cooler surface. Your body temperature at 98.6°F would be much cooler than the automobile parked in the 120°F heat, and the law of heat transfer dictates that your body would be attempting to absorb heat from the warmer surface.

Figure 1-1 further illustrates the transfer of heat. Pouring one quart of liquid at 90°F into another quart of liquid at 70°F results in a two-quart container at 80°F.

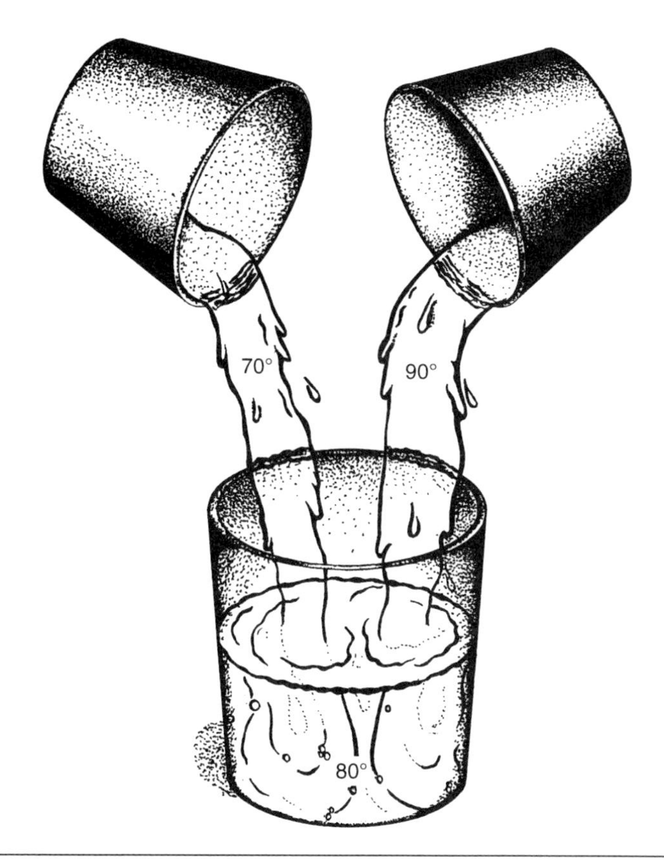

Figure 1-1 Heat always moves from a warmer surface to a cooler surface, and in the event that the volume of the two substances are equal, an equalization of temperature occurs. Pouring a one-quart container at 70°F and a one-quart container at 90°F into a two-quart container results in two quarts of liquid at 80°F.

QUICK NOTE

Another law of thermodynamics is that heat moves in three ways: radiation, conduction, and convection.

RADIATION

Radiant heat travels from warmer objects to cooler objects but does not heat the space in between. The best illustration of heat movement through radiation is the heat from the sun. Although the sun is 93,000,000 miles away, it will warm you, or a building, or a concrete surface, yet it does not warm the atmosphere 5,000 miles above the earth's surface.

Figure 1-2 Infrared heating units are an illustration of heat movement through radiation. An overhead unit, such as the one shown, is used in drafty work areas to warm workers without warming the air.

Gas-fired or electric infrared heating units such as the one shown in Figure 1-2 are another illustration of heat transfer through radiation.

CONDUCTION

Conduction is defined as heat transfer through a substance. In a refrigeration system, heat is transferred via conduction through the metal (usually copper or aluminum) coil that contains the refrigerant. An example of conduction is illustrated in Figure 1-3. The heat from the flame travels through the copper tubing, which soon becomes too hot to handle.

CONVECTION

The third method of heat transfer is convection. Convection is the transfer of heat within a fluid (liquid or gas) from a higher density area to

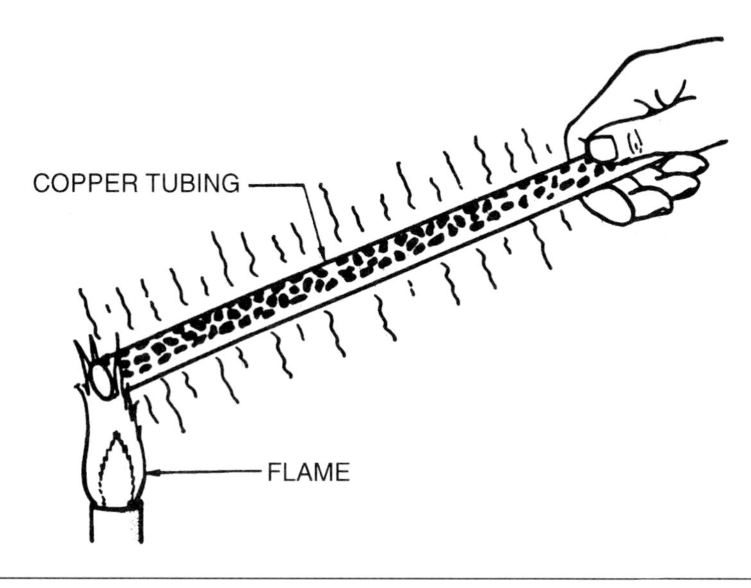

Figure 1-3 Conduction can be illustrated by placing one end of a length of copper tubing into an open flame. In time, the heat will travel the full length of the tubing.

a lower density area. For example, in a pan of water on a stove, the water at the bottom of the pan is heated first. This warmer water tends to rise, while the cooler water at the top tends to sink. This creates a constant flow from top to bottom until all of the water is heated to the same temperature.

THE BRITISH THERMAL UNIT

The British Thermal Unit (BTU) is defined as the amount of heat required to raise the temperature of one pound of water one degree Fahrenheit. When applied to refrigeration, it refers to the removal of a given amount of heat, resulting in a drop in temperature.

There are 12,000 BTU's in a ton of refrigeration capacity. This means that a one-ton refrigeration system is capable of removing 12,000 BTU's of heat per hour, a two-ton unit can remove 24,000 BTU's per hour, a three-ton unit 36,000 BTU's, and so on.

UNDERSTANDING SENSIBLE AND LATENT HEAT

When a refrigeration system operates, it removes two types of heat: sensible heat and latent heat.

Sensible Heat

Sensible heat is the easiest to understand, and is defined as heat that can be measured, and that brings about a change in temperature. When you put a thermometer in your mouth and the temperature rises, the thermometer is measuring an increase in sensible heat.

Latent Heat

Latent heat, also referred to as *hidden heat*, is sometimes more difficult to understand. Latent heat is heat that brings about a change in state, but does not bring about a change in temperature.

To understand latent heat, consider the change of state from water to ice (Figure 1-4). As heat is removed from the water and the water is chilled to 32°F, it begins to change state from water to ice. It stays at 32°F until the change of state is complete. During the time that it remains at 32°F, a great deal of heat is being removed. This is known as *latent heat*. When you are able to measure the temperature of the ice at 31°F, you are again dealing with sensible heat, the removal of which brings about a change in temperature.

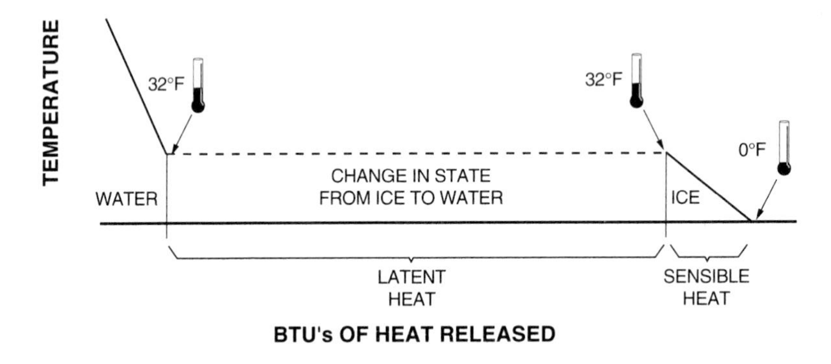

Figure 1-4 Latent heat brings about a change in state, but not a change in temperature.

UNIT ONE SUMMARY

In order to be an effective troubleshooter, you need an understanding of refrigeration fundamentals. Refrigeration is the transfer of heat from a place where it is not wanted to a place where it is unobjectionable. A refrigeration system does not put cold air in, it takes the heat out.

The laws of heat transfer, also known as the laws of thermodynamics, apply to all refrigeration systems. One law of thermodynamics states that heat always moves from a warmer surface to a cooler surface. Another law states that heat moves in three ways:

1. Radiation, which is the movement of heat from a warmer object to a cooler object without heating the space in between.
2. Conduction, which is the movement of heat through a substance.
3. Convection, which is the transfer of heat in a fluid from a higher density area to a lower density area.

Refrigeration capacity is measured using the British Thermal Unit (BTU). It is defined as the amount of energy required to raise one pound of water one degree Fahrenheit. There are 12,000 BTU's per ton of refrigeration capacity.

The two types of heat removed by a refrigeration system are sensible heat, which brings about a change in temperature, and latent heat, which brings about a change in state but not a change in temperature.

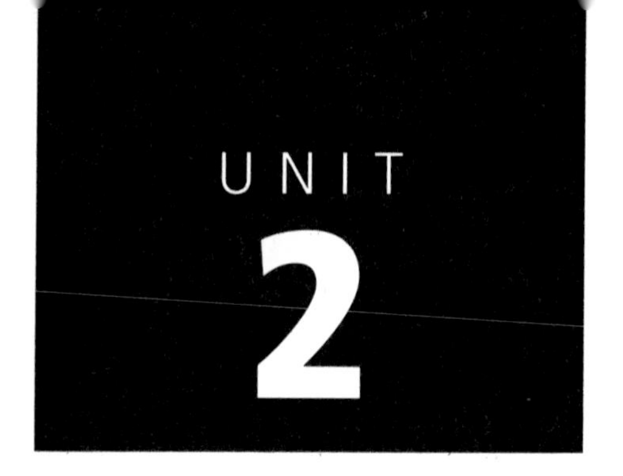

System Components and Accessories

All refrigeration systems have four basic components: compressor, condenser, evaporator, and metering device. When a refrigeration system is designed for a specific application, various accessories may also be used. These include filter driers, accumulators, and receivers. This unit provides an overview of these basic system components and accessories.

COMPRESSORS

The compressor in a refrigeration system works as a vapor pump. It accepts a low pressure gas on one side and discharges a high pressure gas on the other side. When the refrigerant in a system reaches the compressor, it must be in the vapor state. In the event that liquid does get to the compressor, a condition known as *slugging* occurs and the compressor may be damaged because it is trying to compress something that cannot be compressed—a liquid.

There are five basic categories of compressors: reciprocating (also known as a *piston compressor*), rotary, screw, centrifugal, and scroll. Within some of these categories there are more than one specific type of compressor. We will cover the reciprocating compressor first.

Reciprocating Compressors

The inside of a reciprocating compressor appears somewhat like the cylinder/piston/crankshaft assembly of an automobile engine. Its assigned task, however, is much different than that of an automobile engine. (See Figure 2-1.)

QUICK NOTE

A fundamental law of physics states that if you reduce the area occupied by a gas, the pressure of the gas will increase.

A reciprocating compressor can be driven in several different ways. It may be an open-drive compressor (belt-driven or coupler-driven) or it may be driven with an integral motor (hermetic and semi-hermetic types).

Open-Drive Reciprocating Compressors – Open-drive compressors include the belt-driven and coupler-driven types. Belt-driven compressors are found on older units. A setup such as this would be similar in appearance to an air compressor unit such as you might see in a garage or paint shop (minus the air storage tank). Belt-driven compressors may still be found in neighborhood grocery stores or taverns in which the equipment has been in service for many years.

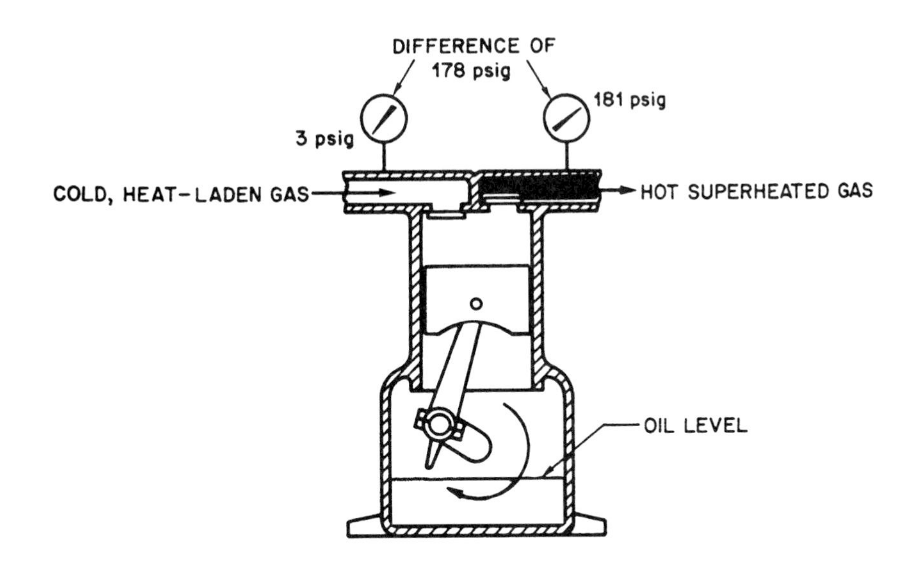

Figure 2-1 In a reciprocating compressor, the temperature and pressure of the refrigerant are increased when the area occupied by the vapor is decreased by the action of the piston.

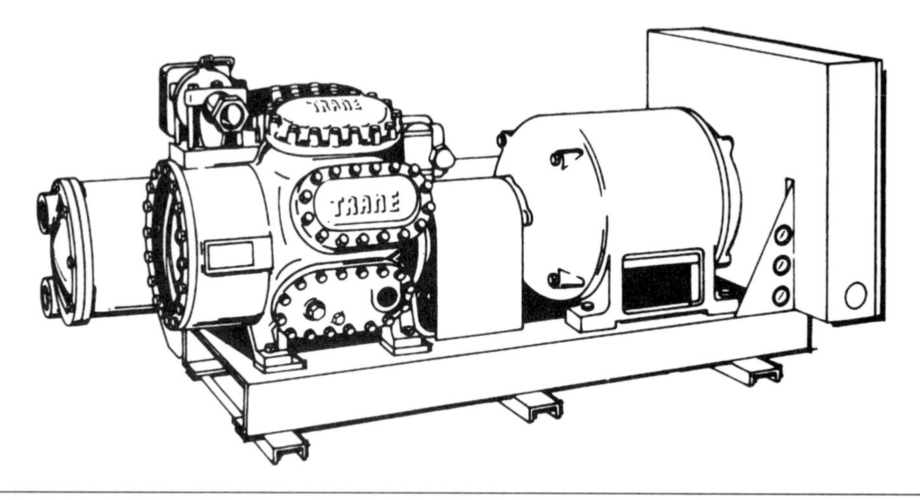

Figure 2-2 On larger chiller systems such as those found in apartment complexes and office buildings, a coupler-driven compressor is often used. (Courtesy The Trane Company, La Crosse, WI)

A second method of operating an open-drive compressor is a coupler-driven system in which the electric motor, instead of sitting next to the compressor, faces the compressor shaft-to-shaft and a coupler is used to connect the two. Coupler-driven compressors are common on large chiller systems found in office buildings or apartment complexes. An example of a coupler-driven, open-drive compressor is shown in Figure 2-2.

Integral Reciprocating Compressors – These compressors have an electric motor as an integral part of the compressor itself. In other words, the motor is built into the compressor rather than being a separate component that is connected externally. There are two types of integral reciprocating compressors popularly used in HVAC-R equipment: the hermetic or *welded-dome compressor*, and the semi-hermetic compressor (sometimes referred to as a *serviceable compressor*). The components within a hermetic compressor are shown in Figure 2-3. A semi-hermetic compressor is shown in Figure 2-4.

Hermetic compressors are hermetically sealed and cannot be field-serviced. They are found on many different types of units including refrigerators and freezers, some walk-in coolers and freezers, self-contained ice cream freezers such as those used in convenience stores, commercial ice machines, and comfort cooling units up to 7½ tons of capacity. A unit over that size would have a semi-hermetic compressor.

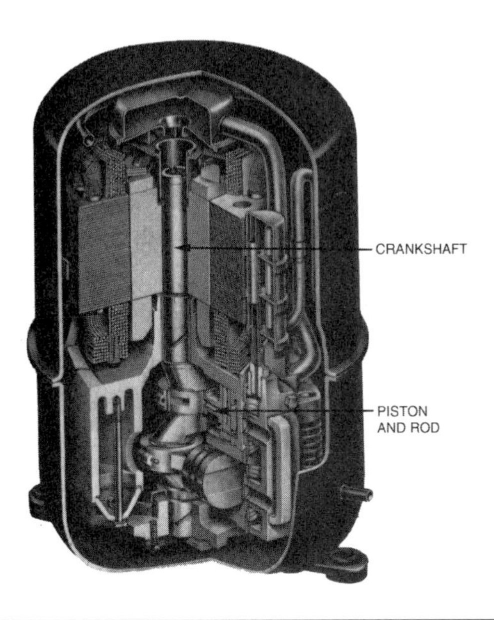

Figure 2-3 A welded-dome compressor is also known as a hermetic compressor and is not serviceable. (Courtesy Tecumseh Products Company)

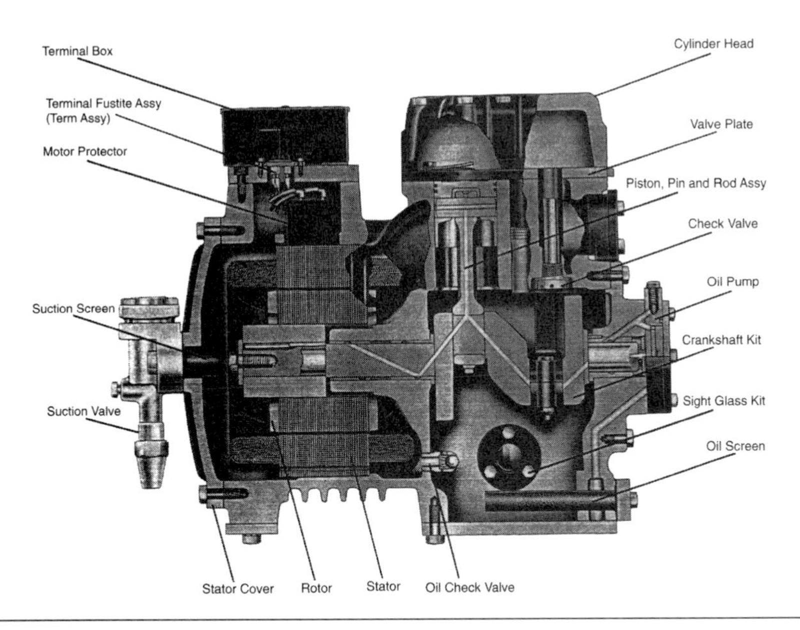

Figure 2-4 A semi-hermetic compressor can be disassembled for some field service repairs. Specialty shops remanufacture semi-hermetic compressors. (Courtesy Copeland Corporation)

Semi-hermetic reciprocating compressors include all those over 7½ tons as well as some smaller units. For example, semi-hermetic units as small as ¾-ton may be found in reach-in refrigerators such as those in restaurants. It is also common to find these compressors in grocery store equipment rooms. You can remove the cylinder head in a semi-hermetic compressor and accomplish valve or valve plate replacement in the field. A semi-hermetic compressor can also be completely rebuilt or remanufactured in specialty shops.

QUICK NOTE

Reciprocating compressors, whether they are hermetic, semi-hermetic, or open-drive, are cooled by suction gas as it returns to the compressor dome.

Rotary Compressors

Rotary compressors are a type of hermetic compressor commonly used in domestic refrigerators and window air conditioners. A rotary compressor can be one of two classifications: fixed vane or rotating vane. Both types are shown in Figure 2-5.

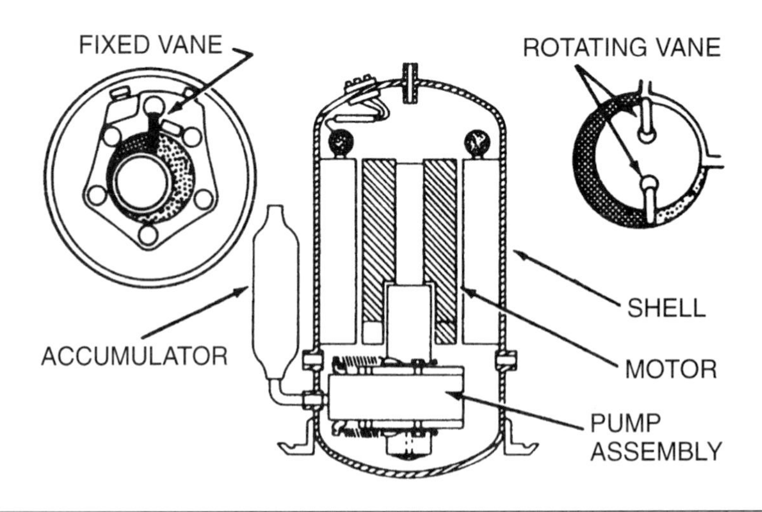

Figure 2-5 Two types of rotary compressors are the fixed vane and the rotating vane. (Courtesy Fedders North America)

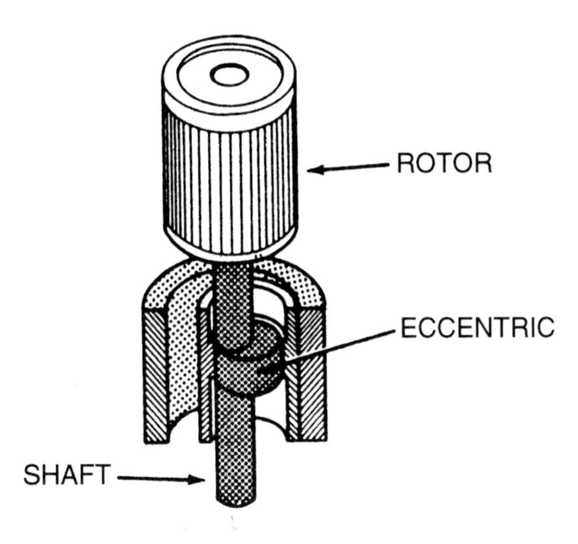

Figure 2-6 The eccentric is the component that creates the elliptical pattern in a rotary compressor. (Courtesy Fedders North America)

When used in window air conditioners, the rotary compressor such as the one shown in Figure 2-5 comes equipped with an accumulator. This is one case in which an accessory is an integral part of the component. (Suction line accumulators are discussed later in this unit.)

A rotary compressor reduces the area occupied by a vapor through the use of what is known as an *eccentric*. This is a component of the rotary compressor that follows an elliptical pattern rather than a perfect circle. As the eccentric rotates, the low pressure vapor is compressed and exits the compressor as a high pressure vapor. An example of the eccentric is shown in Figure 2-6 and the pattern it creates is shown in sequence in Figure 2-7.

To understand the operation of a rotary compressor, follow the sequence of the eccentric shown in Figure 2-7. The arrows indicate refrigerant coming into the compressor (suction on the left side of the eccentric) and exiting the compressor (discharge on the right side of the eccentric). In a rotary compressor, suction and discharge take place simultaneously.

In the first diagram in Figure 2-7, the eccentric is shown at 50 percent suction. As the eccentric rotates as shown in the second diagram at 99 percent suction, one complete revolution has been accomplished. The third diagram shows the eccentric at 1 percent suction, and in the final diagram, the sequence of rotation is shown at 25 percent suction.

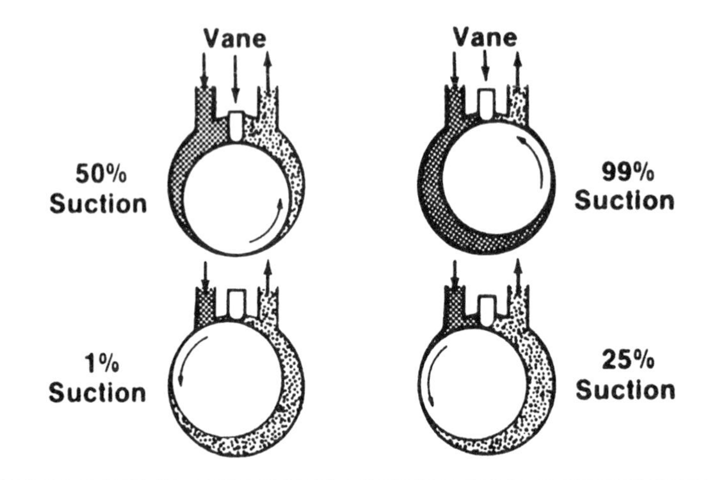

Figure 2-7 In a rotary compressor, suction and discharge take place simultaneously. (Courtesy Fedders North America)

QUICK NOTE

A rotary compressor differs from a reciprocating compressor in that it is cooled by the discharge gas rather than the suction gas.

Screw Compressors

A screw compressor uses two gear-like assemblies that are machined to fit closely together. As these gears rotate, they create a suction and pressure rise in a constant manner similar to the operation of a rotary compressor. The sequence of rotation of a screw compressor assembly is shown in Figure 2-8.

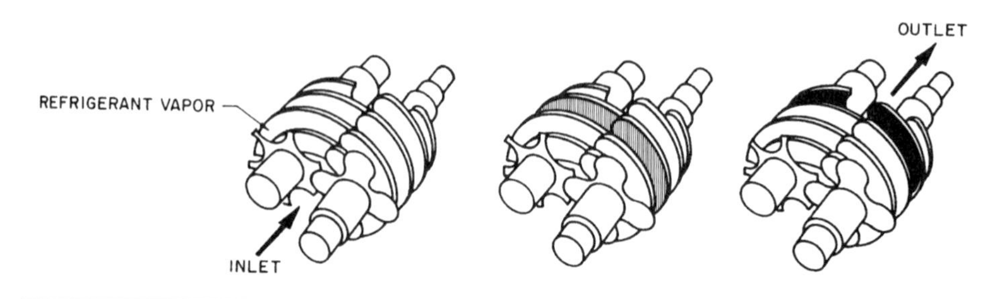

Figure 2-8 A screw compressor uses two closely machined components to create the pressure rise in a refrigeration system.

Screw compressors are commonly found in commercial and industrial refrigeration applications. They are usually found as an open-drive type of unit, but hermetic screw compressors may occasionally be found in a capacity up to 7½ tons.

Centrifugal Compressors

Centrifugal compressors are common in large refrigeration systems used to chill water for comfort cooling systems in apartment complexes and office buildings. The inner components of a centrifugal unit are similar to the impeller in a pump. The impeller rotates, allowing suction into the housing, and as it continues to rotate, it discharges the refrigerant at a slightly higher pressure than it was at suction intake.

Scroll Compressors

The use of scroll compressors in comfort cooling units is increasing. This type of compressor uses two scrolls fitted closely together. As an orbiting action is accomplished, the scroll system allows for suction at a low pressure and discharge at a higher pressure. The suction and discharge sequence of a scroll compressor is shown in Figure 2-9.

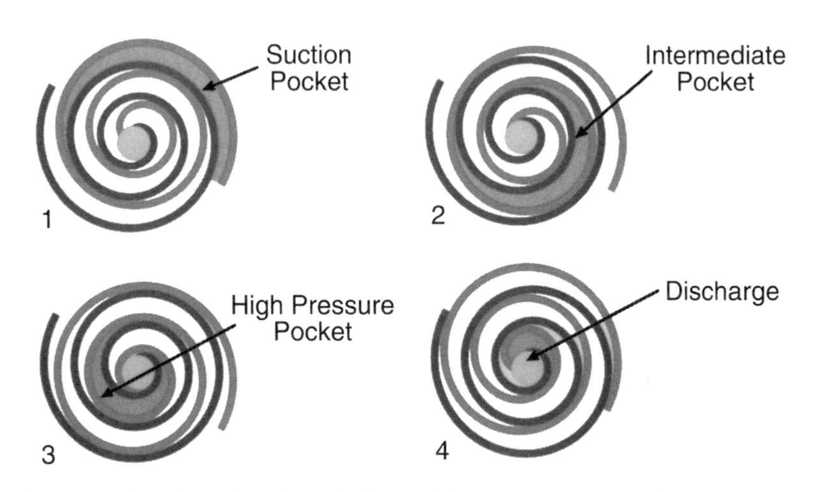

Figure 2-9 The sequence of operation of a scroll compressor. (Courtesy Copeland Corporation)

CONDENSERS

The refrigeration system condenser is a heat exchanger in which the compressed refrigerant vapor is cooled until it becomes a liquid. Common types of condensers include static, forced-air, and water-cooled.

Static Condensers

A static condenser allows the rejection of heat simply through natural air flow without the assistance of a forced-air system or the use of water as a heat transfer agent. Static condensers are most commonly found on certain models of domestic refrigerators and freezers. They are mounted on the rear of the cabinet, with the tubing itself held in position by welded wires.

Forced-Air Condensers

Forced-air or *fan-cooled condensers* are also used on some domestic refrigeration units. In addition, they are used in comfort cooling systems and room air conditioners, as well as in many commercial refrigeration applications, such as walk-in coolers, reach-in refrigerators, and ice machines. A forced-air condenser differs in appearance from the static type in that the tubing is shaped to form a more compact "bundle" in which cooling fins support the tubing and allow air flow across the tubing in an even pattern. An example of a forced-air condenser in a window air conditioner is shown in Figure 2-10.

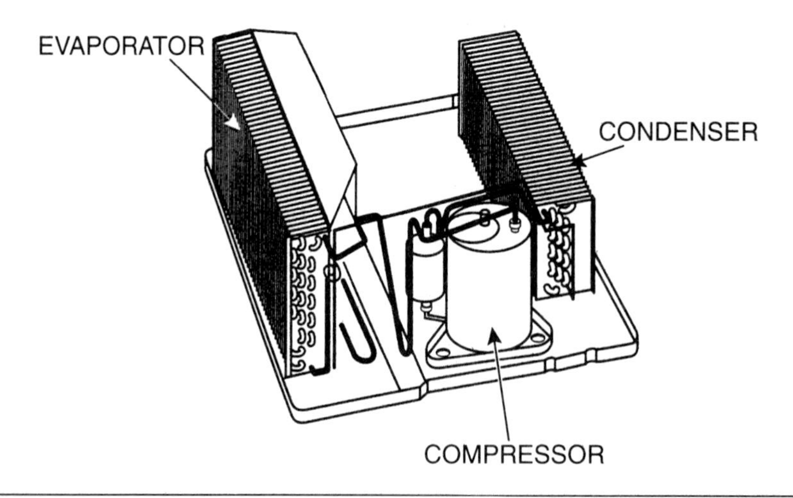

Figure 2-10 A finned tubing assembly is found in window air conditioners.

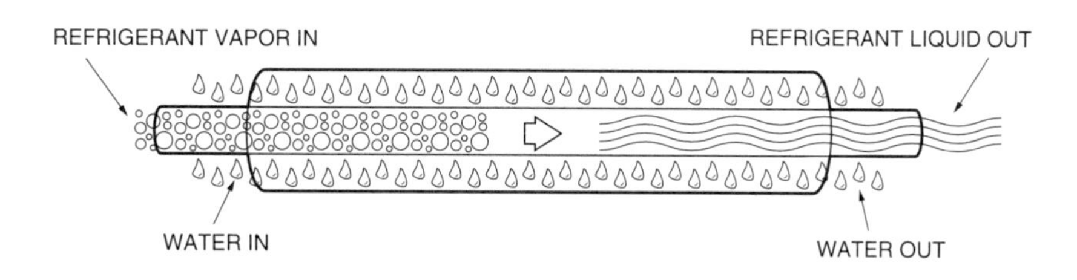

Figure 2-11 A tube-in-tube condenser allows the use of water as a heat transfer agent in a refrigeration system.

Water-Cooled Condensers

Water-cooled condensers are designed in a variety of ways, depending on the application. Common types include tube-in-tube, shell-and-coil, shell-and-tube, and evaporative. Many water-cooled condensers use valves to adjust the water flow to meet the changing demand.

Tube-in-Tube Condensers – Tube-in-tube condensers are used on small commercial units such as ice machines and some walk-in coolers. These condensers are constructed to allow heat transfer from the refrigerant to the water, but the two substances do not come into direct contact with each other. A condenser of this type can be comparatively small since water has excellent heat transfer characteristics. The basic method of construction of a tube-in-tube condenser is shown in Figure 2-11.

Shell-and-Coil Condensers – Shell-and-coil condensers are found on some commercial refrigeration systems. A shell-and-coil unit works on the same principle as a tube-in-tube unit, with two separate sections. The difference is that a dome-like assembly houses the tubing. (See Figure 2-12.)

Shell-and-Tube Condensers – Shell-and-tube condensers are found on large refrigeration systems used in office buildings and apartment complexes. Shell-and-tube condensers, unlike most tube-in-tube and all shell-and-coil condensers, can be disassembled for cleaning. Specialty tools are used to keep this type of condenser operating properly.

Evaporative Condensers – Evaporative condensers may be found in some commercial cooling systems, such as grocery store equipment rooms. In this type of system, the condenser tubing from all units in the store is piped through a central cooling tower and water is sprayed directly onto the condenser tubing while a fan blows a large volume of air across the tubing. As the water evaporates, it cools the tubing.

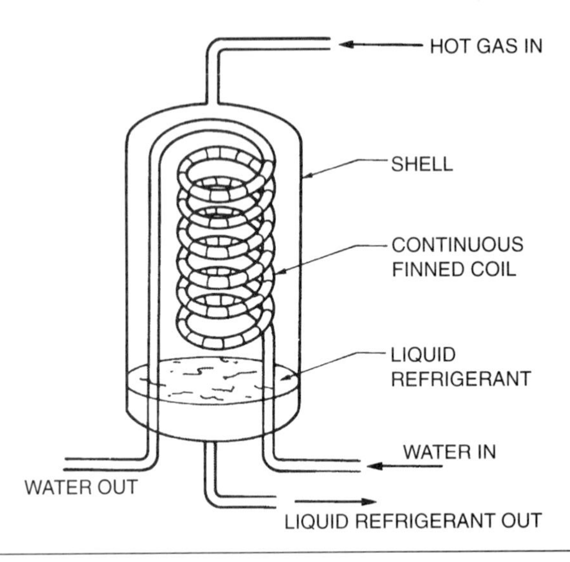

Figure 2-12 In a shell-and-coil condenser assembly, the finned condenser is housed in a welded dome. Water is piped through the dome to cool the hot vapor.

Recirculating Condenser Water – Using water as a heat transfer agent in cooling condensers brings up another issue: Do we just run water past the refrigerant tubing and down the drain to carry away the heat or do we recirculate the water? Some systems use the water, then dump it down the drain. However, most systems recirculate the water so it can be used over and over rather than wasted. A common method of cooling the warmed water is by piping it in a closed loop through a cooling tower that uses a large fan and water in an evaporative method to cool the condenser water piping. This creates a second, independent water system.

TROUBLESHOOTING HINT

If heat transfer is affected either by dirty fins (fan-cooled condenser) or corrosion (water-cooled condenser), then the overall performance of the unit will suffer. You must determine if the customer's complaint of "not cooling" is created by a dirty or corroded condenser and, if so, what to do about it. Air-cooled condensers can be cleaned with commercially-available coil cleaning chemicals sprayed onto the fin and tubing assembly. Some water-cooled condensers may be cleaned with a brush assembly, while others use a chemical distribution system to feed chemicals into the water system on a constant basis.

EVAPORATORS

The purpose of the evaporator is to vaporize the liquid refrigerant and absorb heat into the system. Two common evaporators are the plate-type and the finned-type.

Plate-Type Evaporators

Plate-type evaporators are commonly found in low-load applications, such as domestic refrigerators. The evaporator in this type of unit is shaped to form the freezer section and a separate plate is used in the fresh food section. There is no forced-air system in this case.

Finned-Type Evaporators

Finned-type evaporators use a forced-air system to provide greater capacity. They are found in some refrigerators, and are always used in air-to-air comfort cooling systems, heat pumps, commercial units (such as display cases in grocery stores), walk-in coolers, and window air conditioners. (Refer again to Figure 2-10.) With a finned-type evaporator, a fan is used to increase the volume of air across the coil, which in turn increases the amount of work done by the refrigeration system.

An evaporator can also be found in a system similar in appearance to a shell-and-tube condenser unit. In this case, the water circulated through the chiller barrel assembly is cooled by the refrigerant. The chilled water is then piped through coil assemblies, and air handling systems force air through the coils, cooling the conditioned space.

TROUBLESHOOTING HINT

An evaporator of this type is subject to the same corrosion problems that plague water-cooled condensers. This must be taken into consideration when tracking down a complaint of "not cooling" or when operating costs are higher than normal.

METERING DEVICES

A metering device creates a controlled restriction in the refrigerant path and allows the proper amount of refrigerant to enter the evaporator. It is sometimes referred to as an *expansion device*. There are two common types of metering devices: the capillary tube and the expansion valve.

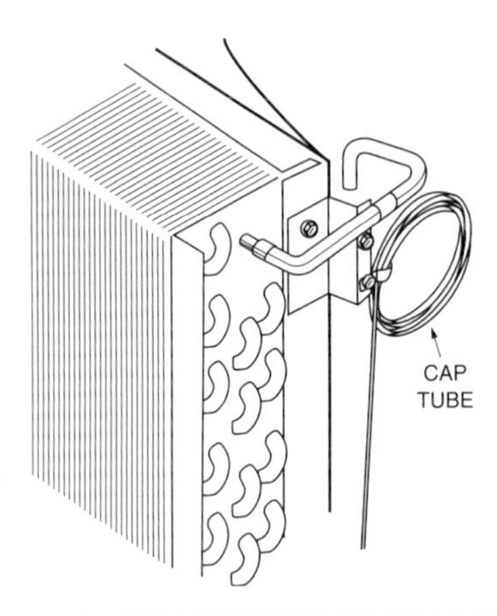

Figure 2-13 A capillary tube is usually coiled up due to its length.

Capillary Tubes

A capillary tube is a length of tubing with a very small inner diameter. It allows the same volume of refrigerant into the evaporator regardless of the load. Capillary or *cap tube systems*, as they are called, may be a single capillary line on small units while larger comfort cooling systems may use multiple capillary tubes. A capillary tube is usually found carefully coiled up because it may be several feet long, even in smaller units (Figure 2-13).

A capillary tube will also be found as a portion of a heat exchanger assembly in domestic refrigerators. In this instance, the capillary tube is soldered to the suction line of the refrigeration system and gives up some of its heat to the cooler suction line as the refrigerant circulates through the system.

Expansion Valves

There are three types of expansion valves used in refrigeration systems: the thermostatic expansion valve, the automatic expansion valve, and the electronic expansion valve.

Thermostatic Expansion Valves – A thermostatic expansion valve (referred to as a TEV or TXV), is commonly found in commercial refrigeration systems such as walk-in coolers and freezers, grocery store equipment, large chiller systems, and on some older residential comfort cooling systems. An example of a TEV is shown in Figure 2-14.

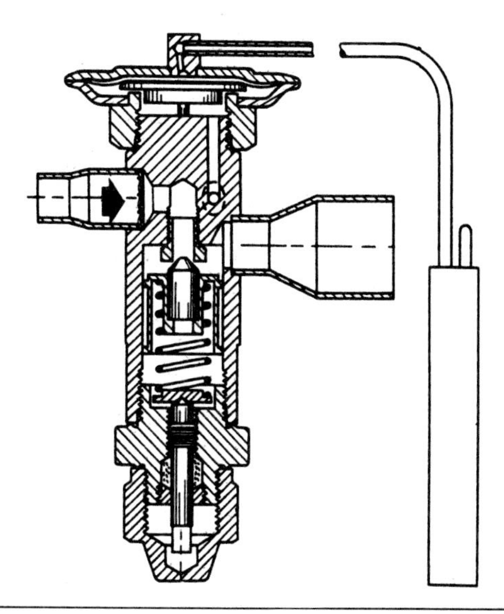

Figure 2-14 A thermostatic expansion valve is a metering device that adjusts the refrigerant flow to meet the system load. (Courtesy Sporlan Valve Company)

A TEV allows a large volume of refrigerant to enter at its inlet, but through the use of its inner components, allows a smaller volume of refrigerant to exit the valve body. It operates on three pressures:

1. Bulb pressure, which is supplied by the sensing bulb attached to the end of the evaporator tubing and tends to cause the valve to open.
2. Spring pressure, which is supplied by the spring inside the valve and tends to cause the valve to close.
3. Evaporator pressure, which is the pressure inside the refrigeration system and also offers some assistance in the closing action of the valve.

The TEV differs from the capillary tube in that it adjusts and allows more or less refrigerant to enter the evaporator as the heat load fluctuates within the conditioned space. When the TEV's sensing bulb detects a rise in temperature, it applies pressure that causes the valve to open and allow more refrigerant to enter the evaporator. As the temperature of the suction line drops, the sensing bulb reacts by applying less pressure and allowing a smaller volume of refrigerant to enter the evaporator. A typical application of a thermostatic expansion valve is shown in Figure 2-15. The TEV is adjusted according to specific service procedures in order to offer the most efficient operation of the refrigeration system.

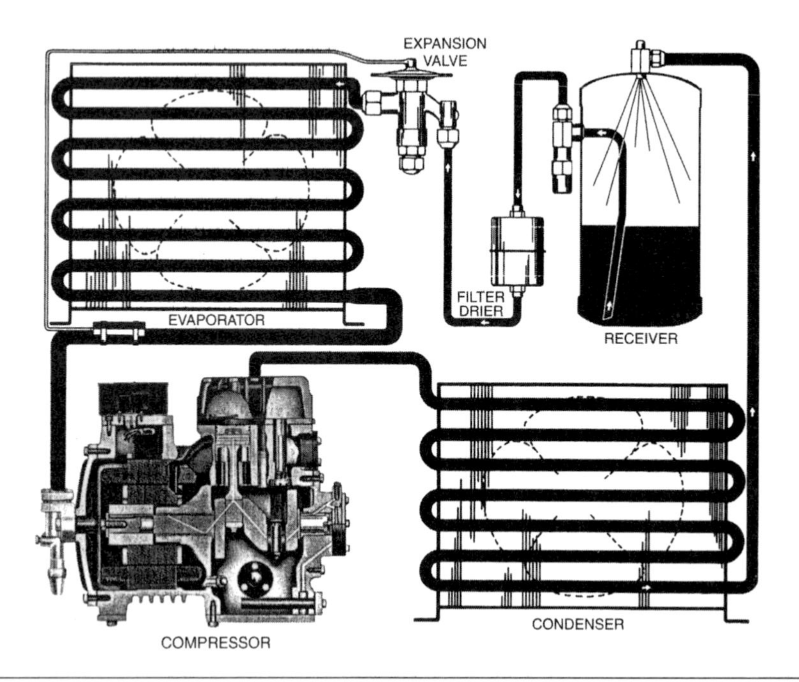

Figure 2-15 A thermostatic expansion valve is typically used in commercial refrigeration systems such as those found in grocery stores or restaurants. (Courtesy Copeland Corporation)

Automatic Expansion Valves – An automatic expansion valve or AEV differs from a TEV in that it uses no sensing bulb. Instead, the AEV allows a constant volume of refrigerant flow and is used in applications such as water fountains where the load remains fairly constant. The AEV can be adjusted for optimum operation.

Electronic Expansion Valves – The electronic expansion valve or EEV is used to offer more precise control of refrigerant into the evaporator. The valve body mounts in the same location as a standard TEV, but instead of a sensing bulb, the EEV uses a thermistor that is positioned directly in the refrigerant flow in the suction line and senses the rise and fall of refrigerant temperature.

The EEV is controlled by changes in the voltage level. Applying more voltage causes the valve to open, while applying less voltage causes the valve to close. The thermistor is a solid state device that is wired in conjunction with the valve. As the temperature rises, the resistance of the thermistor decreases and causes the EEV to allow more refrigerant to flow. As the temperature drops, the thermistor's resistance increases, causing the EEV to allow less refrigerant to flow.

REFRIGERATION SYSTEM ACCESSORIES

In some refrigeration systems, accessories are necessary for optimum performance. Common accessories include the filter drier, accumulator, and receiver. A wide variety of valves and solenoid components are also used, depending on the specific application of the refrigeration system. We will focus on the most common accessories in this unit, with explanations of other accessories in later units on troubleshooting systems.

Filter Driers

The filter drier protects the compressor from damage caused by the introduction of moisture, acids, and other contaminants.

The liquid line filter drier is the most common. In some cases, a filter drier is also installed in the suction line of the system. Some filter driers are specifically designed to filter out wax or acids. Whenever a refrigeration system is serviced, or a component such as a compressor is replaced, the filter drier must be replaced.

Suction Line Accumulators

An accumulator is a storage tank located on the suction side of the refrigeration system. Its function is to protect the compressor from liquid refrigerant that would damage the compressor valves or other components within the compressor.

Liquid Line Receivers

A receiver is also a storage tank for refrigerant. It is located on the discharge side of the refrigeration system, and is commonly found on systems that use a TEV as the metering device. Since the TEV system adjusts the amount of refrigerant to meet the load, the surplus refrigerant is stored in the receiver until the TEV calls for more volume.

UNIT TWO SUMMARY

In any refrigeration system, there are four basic components: the compressor, condenser, evaporator, and metering device. Depending on the specific application of the refrigeration system, accessories may be added to achieve optimum performance.

The five types of compressors are the reciprocating, rotary, screw, centrifugal, and scroll. Some units are hermetically sealed and cannot be serviced. A semi-hermetic or serviceable compressor can be serviced in the field by a technician. Serviceable compressors can also be remanufactured in a specialty shop.

Condensers are separated into two categories, air-cooled and water-cooled. A domestic refrigerator is usually equipped with a static condenser (not forced-air) but may also be equipped with a fan-cooled condenser. Forced-air condensers are most often found on comfort cooling and commercial refrigeration systems. Water-cooled condensers are used on commercial equipment such as ice machines and walk-in coolers, as well as large central cooling systems. The four types of water-cooled condensers are the tube-in-tube, shell-and-coil, shell-and-tube, and evaporative.

Evaporators include the plate-type, which is generally used in domestic refrigerators, and the finned-type, which is also used in some refrigerators as well as all comfort cooling systems and commercial units. An evaporator immersed in water is used on a chiller system to chill the water before pumping it to the air handling coils.

Two types of refrigeration system metering devices are the capillary tube and the expansion valve. Single capillary tube systems are used on small refrigeration systems, while multiple capillary tubes may be found on larger comfort cooling systems. In some cases, the capillary tube is also a part of a heat exchanger system. The thermostatic expansion valve uses a sensing bulb to vary the amount of refrigerant entering the evaporator. An automatic expansion valve is used in applications such as a water cooler. An electronic expansion valve works in tandem with a thermistor that monitors refrigerant temperature. The changing resistance of the thermistor causes the EEV to allow more or less refrigerant into the evaporator.

The most common accessories on a refrigeration system are the filter drier, suction line accumulator, and liquid line receiver. A filter drier is used to keep a system free of moisture, acid, and other contaminants that could damage the compressor. A suction line accumulator is a storage tank that prevents liquid refrigerant from getting to the compressor. A liquid line receiver is commonly used to store excess refrigerant in a system that uses a thermostatic expansion valve as a metering device.

The Refrigeration Cycle

Now that we have identified and reviewed the operation of the fundamental components of a refrigeration system, we will proceed with an explanation of the refrigeration cycle. As a troubleshooting technician, you must understand the method by which refrigeration is accomplished. Technicians with many years of field experience have been known to make costly diagnostic errors due to a lack of understanding of this basic process.

We will explain the refrigeration cycle using the simplified refrigeration system in Figure 3-1.

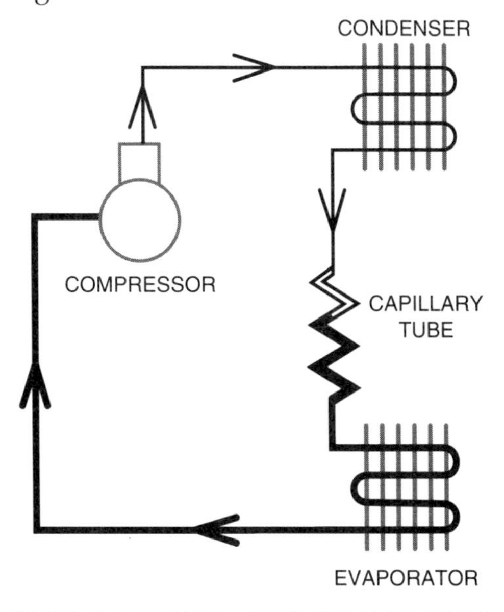

Figure 3-1 Simplified refrigeration system that uses a capillary tube as the metering device.

It shows only the four basic components: compressor, condenser, evaporator, and metering device. In this case, the metering device is a capillary tube, the simplest of all metering devices since it accomplishes its task through only its method of construction and does so on a constant feed basis.

In unit one, we discussed some of the fundamental laws of thermodynamics *(heat always moves from a warmer surface to a cooler surface and heat moves in three ways: conduction, convection, and radiation)*. At this point we will add two additional heat transfer principles.

QUICK NOTE

1. When a substance boils (evaporates), it absorbs heat.
2. When a substance condenses, it rejects heat.

In a refrigeration system, the refrigerant undergoes a change in pressure from one side of the system to the other. It also undergoes a change in state from a liquid to a vapor on the low pressure side of the system, and from a vapor to a liquid on the high pressure side of the system. These changes in state and pressure accomplish the "work" of the system. In other words, when the refrigerant within the system changes from a liquid to a vapor (evaporates) it can perform the work of absorbing heat. On the other side of the system, when the refrigerant changes state from a vapor back to a liquid again (condenses) it can perform the work of rejecting or giving up heat.

For the purpose of explanation, assume that the refrigeration system shown in Figure 3-2 is using refrigerant HFC-134a and that it is located in an ambient temperature of 85°F. The term *ambient* describes the temperature surrounding the system. With this temperature established, we will begin tracing the sequence of the cycle as the refrigerant enters the evaporator. Follow the arrows indicating the direction of refrigerant flow as it starts at the compressor and moves through the condenser and capillary tube. You should be at the point where the capillary tube connects to the evaporator. Note that the evaporator refrigerant temperature is 40°F.

For this example, we will assume that heat has radiated into the conditioned space through windows and an open door. This heat moves toward the evaporator in the system, and since the evaporator (at 40°F) is much cooler than the 85°F outdoor ambient, heat is conducted through the metal tubing of the evaporator and absorbed by the refrigerant.

As the refrigerant in the evaporator absorbs this heat, it begins to boil (evaporate). One of the reasons this process can take place is because

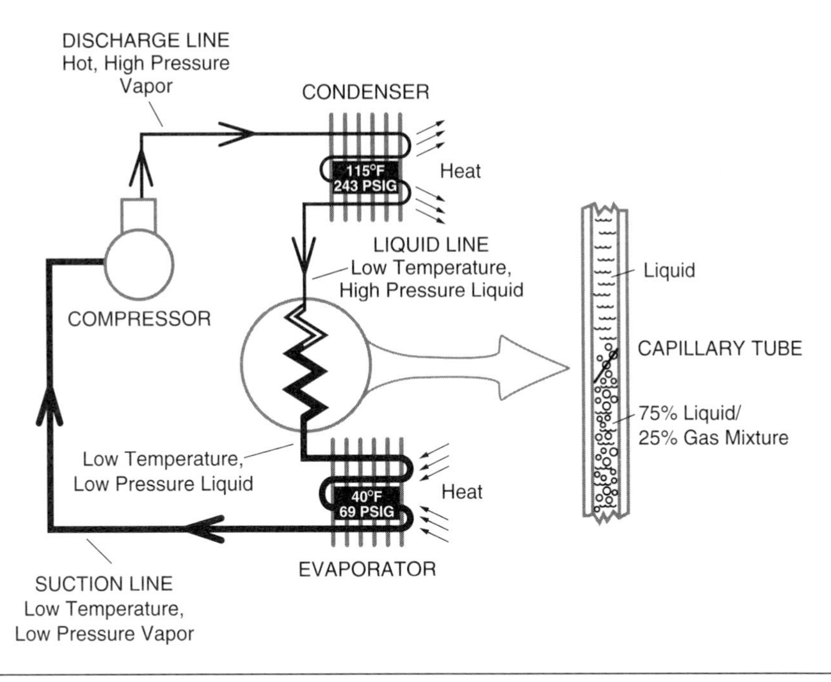

Figure 3-2 Tracking the pressure and temperature of the refrigerant as it flows through the refrigeration system. The thin lines represent the high pressure side of the system and the thick lines represent the low pressure side of the system.

refrigerant is a chemical that has a very low boiling point. For example, HFC-134a (designed as a replacement for R-12), has a boiling point of -15°F at atmospheric pressure (14.7 psi or 0 psig). It is used in refrigerators, freezers, some commercial equipment, and auto air conditioners. (Note: under EPA regulations, R-12 is being phased out because it is believed to be responsible for damage to the ozone layer.) Another common refrigerant, HCFC-22 (formerly known as R-22), has a boiling point of -41°F. HCFC-22 is used extensively in comfort cooling systems.

The chemical makeup of a refrigerant is the reason the action of "boiling" (something we usually relate to a high temperature) can be taking place while at the same time the coil can be chilled, or even freezing.

QUICK NOTE

The boiling point of a substance is directly affected by pressure. As the pressure increases, the boiling point also increases, and vice versa. This means that all refrigerants have a range of temperatures at which they will boil, depending on the pressure at that point in the system.

The evaporator pressure in this system is 69 psig. Because of the increased pressure, HFC-134a will boil at a temperature of 40°F, and the change in state from a liquid to a vapor will continue to take place in the evaporator until the refrigerant has absorbed all the heat it is capable of picking up. (Note: even though the refrigerant in the evaporator is at a much higher pressure than atmospheric, as we will see, it is still at a *low* pressure when compared with the condenser.)

Following the direction of the arrows in the illustration, the refrigerant vapor now travels to the compressor. The tubing that connects the evaporator and compressor is known as the *suction line*.

The compressor accepts the low pressure vapor and compresses it, increasing both the pressure and the temperature. The hot, high pressure vapor travels to the condenser via the *discharge line* and enters the condenser at a temperature that is higher than the ambient temperature. For our purposes, we will assign a condenser refrigerant temperature of 115°F. Believe it or not, because the refrigerant in the condenser is at such an extremely high pressure (around 243 psig) the refrigerant boils at this temperature and will begin to condense as soon as heat is released. Since the refrigerant temperature is higher than the ambient temperature of 85°F, heat loss occurs quickly as heat flows from the condenser into the surrounding air. Within the condenser, the change in state from a vapor to a liquid is occurring. Remember, when a substance changes state from a vapor to a liquid, heat is rejected.

Within the condenser, heat moves through the refrigerant fluid by convection, moves through the condenser tubing by conduction, and is released to the surrounding air by radiation.

At this point, the refrigerant has been cooled but is still under high pressure. The refrigerant leaves the condenser through what is known as the *liquid line* and flows to the metering device, which in our example is the capillary tube. As the refrigerant travels through the small bore of the capillary tube, it creates the pressure drop that is essential to the operation of a refrigeration system. As this occurs, some of the refrigerant "flashes" into a gas, which further cools the rest of the refrigerant. The refrigerant leaves the capillary tube as a low temperature, low pressure liquid/gas mixture.

Upon entering the much larger tubing of the evaporator, there is room for expansion and the refrigerant is allowed to complete the change in state from a liquid to a vapor. The cycle then repeats itself.

HIGH PRESSURE AND LOW PRESSURE

For the purpose of illustration, you could draw a diagonal line through the system that crosses the compressor and the capillary tube

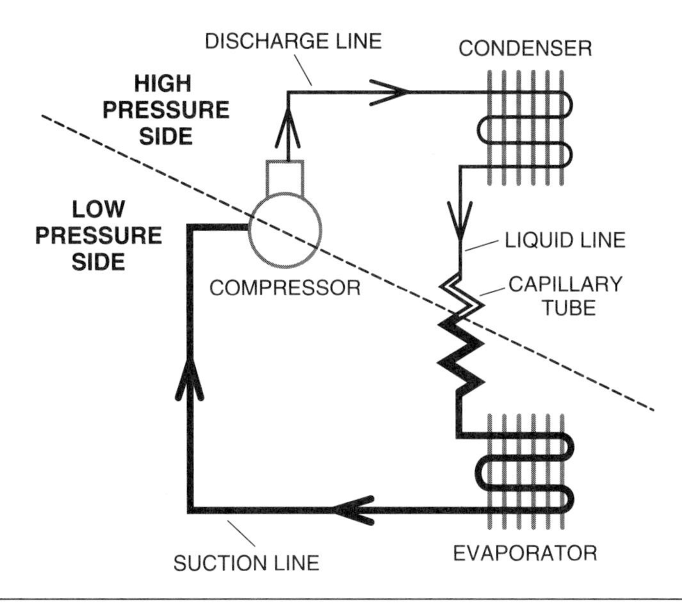

Figure 3-3 Dividing the system into its low pressure and high pressure components. The compressor and metering device create the pressure differential necessary to operate the refrigeration system.

(Figure 3-3). The evaporator, suction line, and compressor dome are considered to be on the low pressure side of the system. The discharge side of the compressor, discharge line, condenser, and liquid line are considered to be on the high pressure side of the system. The compressor and capillary tube are the components of the system that create the pressure differential necessary to accomplish the refrigeration cycle.

PURPOSE OF EVACUATION

Before refrigerant is added to a refrigeration system, the components are evacuated. The purpose of evacuation is to remove all air from the system. This is necessary because air is a non-condensable gas and in order for a refrigeration system to operate as efficiently as possible, none of the space inside the tubing can be occupied by a substance other than refrigerant and the oil that is traveling with it.

The presence of air in a system will also create other problems. Air contains moisture and any moisture in a refrigeration system can damage the compressor and also freeze at the metering device, restricting the refrigerant flow. Air also affects the system operating pressure. This creates problems for the technician who is using gauges to diagnose a problem or monitor system performance.

UNIT THREE SUMMARY

When considering the refrigeration cycle, two additional laws of thermodynamics must be understood: when a substance boils, it absorbs heat, and when a substance condenses, it rejects heat.

A refrigerant is a chemical that has a low boiling point. As a result, a refrigeration system evaporator can contain refrigerant that is boiling while at the same time be cool. HFC-134a (designed as a replacement for R-12) is now used in refrigerators, freezers, some commercial equipment, and auto air conditioners. It has a boiling point of -15°F at atmospheric pressure (14.7 psi or 0 psig). HCFC-22 (formerly known as R-22) is used in comfort cooling systems and has a boiling point of -41°F. The boiling point of a refrigerant is directly affected by pressure.

In the refrigeration cycle, heat is absorbed by the refrigerant in the evaporator because of the change in state from a liquid to a vapor. The refrigerant vapor then flows through the suction line to the compressor, where its temperature and pressure are increased. The hot, high pressure vapor enters the condenser and the change in state from a vapor to a liquid allows the rejection of heat into the surrounding area. The refrigerant leaves the condenser via the liquid line and enters the metering device, which creates a pressure drop. After leaving the metering device, the refrigerant again enters the evaporator as a low pressure liquid/gas mixture and the cycle repeats.

The evaporator, suction line, and compressor dome are considered to be on the low pressure side of the system and the compressor discharge line, condenser, and liquid line are considered to be on the high pressure side of the system.

The purpose of system evacuation is to eliminate moisture and other contaminants. Air is a non-condensable gas and affects the efficiency and operating pressures of a refrigeration system.

SECTION

2

ELECTRICAL FUNDAMENTALS

Like an understanding of refrigeration, an understanding of electricity is a fundamental building block in becoming an effective troubleshooter. To avoid guesswork and ensure safety when servicing HVAC-R equipment, you must understand how alternating current is generated, how it is transported from the generating station to the end user, and how it is distributed throughout a residence or commercial building.

Having a full understanding of electricity will give you the confidence you need to isolate electrical problems within a system and perform an effective repair. Approximately 75 percent of the problems encountered in HVAC-R equipment are electrical in nature.

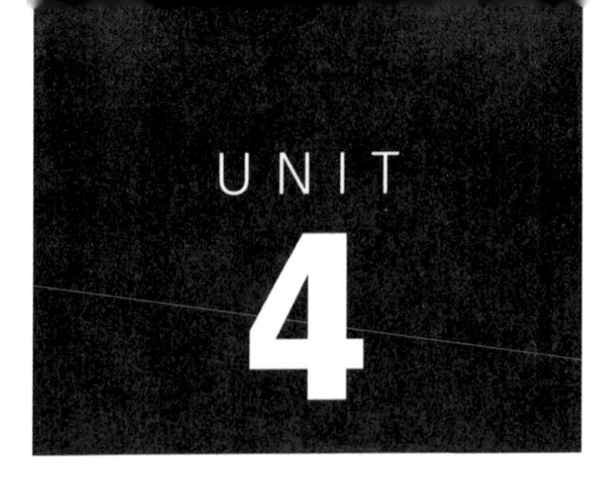

UNIT 4

Alternating Current

The electrical energy used to power HVAC-R equipment found in homes and commercial buildings is alternating current (AC). AC is produced by an alternator and the process by which this occurs can be explained by the theory of electromagnetism.

QUICK NOTE

The theory of electromagnetism, first proposed in the 1800's, states that cutting the lines of force of a magnetic field causes energy to be induced in the conductor. The lines of force of a magnetic field are shown in Figure 4-1.

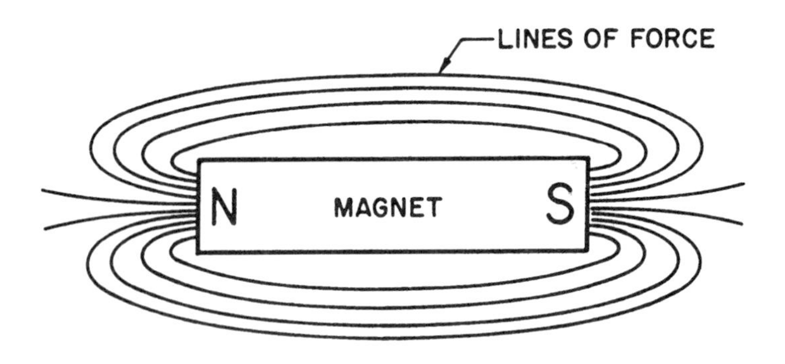

Figure 4-1 Permanent magnet showing lines of force.

Electromagnetism is easily illustrated through the use of magnets. Positioning two magnets close together with their *unlike* poles almost touching causes the magnets to attract each other. Positioning two magnets

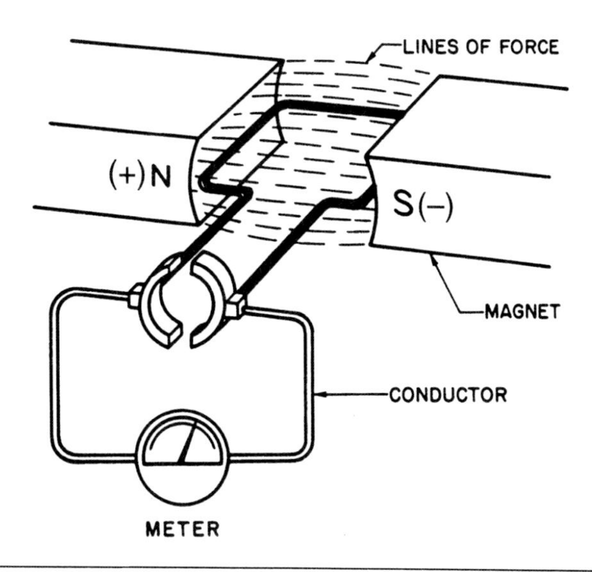

Figure 4-2 Elementary generator.

close together with their *like* poles almost touching causes the magnets to repel, pushing them apart. The pulling or pushing motion demonstrated between the magnets is an example of a magnetic field. There is also a magnetic field surrounding the earth. The needle in a compass points north due to the magnetic "pull" of the earth's magnetic field.

Electricity also produces a magnetic field. In Figure 4-2, the north and south poles of a magnet set up a positive/negative field of energy and the rotation of a conductor induces a current in the conductor.

One cycle is accomplished when the conductor completes a full 360° rotation. Each cycle results in the achievement of a positive peak and a negative peak, which produces two flow reversals. This is illustrated through what is known as a *sine wave* (Figure 4-3). The positive and negative peaks explain the origin of the term *alternating current*. It is the flow of current first in one direction, then in the opposite direction.

As shown, the positive peak is attained when the conductor has rotated from point A to point B, and the positive half of the cycle is achieved when the conductor continues from point B to point C. The negative peak is reached at point D and the negative half of the cycle is accomplished when the conductor returns to point A.

The number of cycles per second or *frequency* of the current is measured in hertz (abbreviated Hz). The term *hertz* is used to honor a nineteenth-century German physicist named Heinrich Rudolph Hertz, who proved the existence of electromagnetic waves.

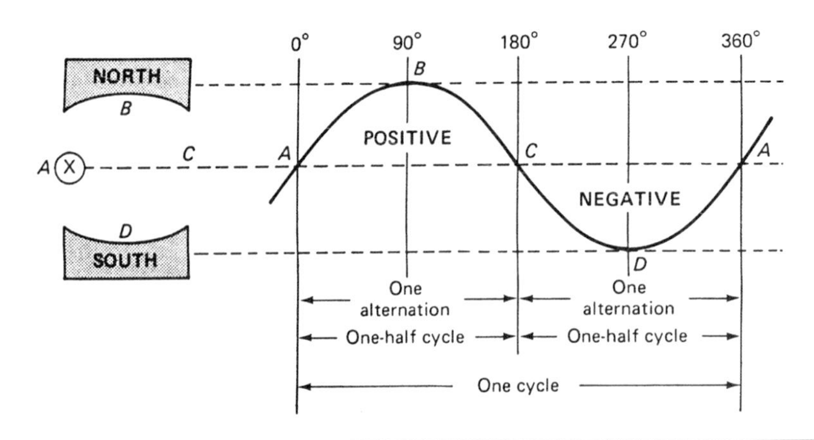

Figure 4-3 A sine wave illustrates the positive/negative concept of alternating current.

In the United States, alternating current is delivered at a frequency of 60 hertz (60 cycles per second). In other words, the rotation we outlined through points A, B, C, and D in Figure 4-3 is accomplished in a time frame of 1/60th of a second. This idea may be easier to understand if you attach the familiar concept of RPM (revolutions per minute) to it. This can be done through some simple multiplication.

Multiplying 60 (the number of revolutions per second) times 60 (the number of seconds in a minute) produces a result of 3,600 revolutions per minute. This means that a 2-pole generator being turned at 3,600 RPM will deliver alternating current at a frequency of 60 hertz.

Rotating a conductor through a magnetic field is not the only method by which electromagnetic current is generated. In many instances, the conductors inside a generator are stationary and the magnets are attached to a rotating assembly. It makes no difference whether the magnets are the stationary or rotating portion of the generator. The effect of cutting the lines of force of the magnetic field is the same—electrical energy is produced.

GENERATING STATIONS

Almost all of the electrical power we use is produced at large generating stations. The energy used to turn the rotating section of a generator commonly comes from heat derived through the burning of fuel, such as oil or coal. This heat travels through a heat exchanger system that contains water. Heating the water creates steam, and when the steam is released in a controlled manner to contact the fins of a turbine, the rotation of the generator rotor is accomplished. Other methods of creating the mechanical energy necessary to turn the turbine may be heat from nuclear fission or water power.

ELECTRICAL DISTRIBUTION SYSTEMS

Once the electrical energy has been generated, the task of electrical distribution must be accomplished. The generating plant commonly delivers electrical energy from its generators and transformers at a very high level. The "electrical pressure" at which electrical energy is delivered, also known as *electromotive force*, is measured in volts. The term *volt* is derived from the name of an Italian professor of physics, Allesandro Volta, who discovered in 1800 that a chemical action between moisture and two dissimilar metals produced electricity.

In order to deliver electrical energy to the end user at an acceptable level, the voltage is "stepped up" or increased by transformers before transmission. This is known as the *transmission voltage*. Substations then use transformers to "step down" the voltage to a lower *distribution voltage*, which is further reduced to the specific *secondary voltages* required by commercial and residential systems. A typical electrical distribution system is shown in Figure 4-4.

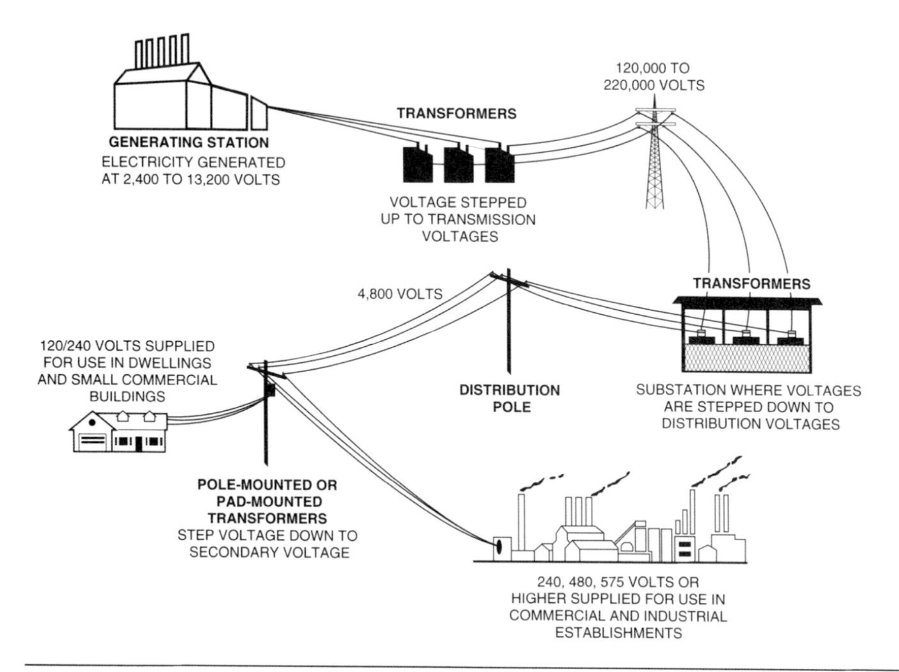

Figure 4-4 Power distribution system.

CONDUCTORS, INSULATORS, AND SEMI-CONDUCTORS

Conductors, insulators, and semi-conductors can be thought of as the "traffic light" in the electrical system. Conductors permit the passage of electricity, insulators halt the passage of electricity, and semi-conductors fit somewhere in between.

Conductors

Once electrical energy is generated, it is transported through conductors. A conductor is a material with a low resistance to current flow. The atomic structure of a given material determines whether or not it is a good conductor. To understand the effect of atomic structure on conductivity, we must first learn the basics of atomic theory.

QUICK NOTE

All substances, whether they are solids, liquids, or gases, are made up of atoms. An atom is the smallest part of an element that can exist alone and still retain all of the properties of the element.

Atoms are electrical in structure. The three principal parts of an elementary atom are shown in Figure 4-5. They are the *electron,* which has a negative charge, the *proton,* which has a positive charge, and the *neutron,* which has no charge. It is considered to be neutral.

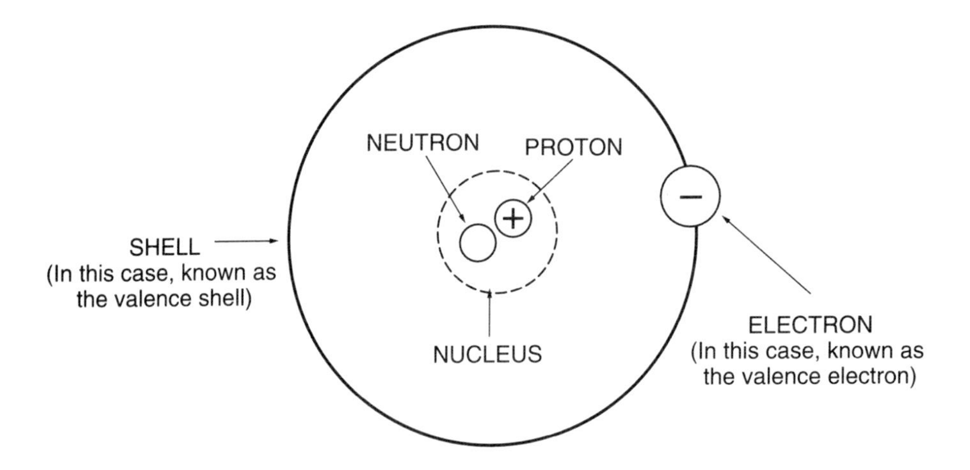

Figure 4-5 Elementary atom.

Neutrons and protons make up the nucleus of the atom. The negatively charged electrons orbit the nucleus. There can be several layers or *shells* of electrons around the nucleus. The electrons in the outermost shell are known as the *valence electrons*.

QUICK NOTE

A basic law of physics, known as the *law of electric charges*, states that opposite charges attract and like charges repel.

Protons and electrons are attracted to each other because they have opposite charges. When an atom contains an equal number of protons and electrons, it is considered to be electrically neutral because of the balance between the negative and positive charges.

An atom will remain neutral (balanced) unless a form of energy is induced in the material. When this occurs, the energy (electricity) can cause an atom to lose an electron. Electricity is the flow of electrons through a material. Free electrons can knock valence electrons out of orbit and, once free, these electrons can knock other valence electrons out of their orbits, and so on. This is the process that allows electrical energy to be conducted along a material until it reaches its destination and performs useful work. Figure 4-6 shows an imbalance being created in the atom, allowing energy to flow through the conductor.

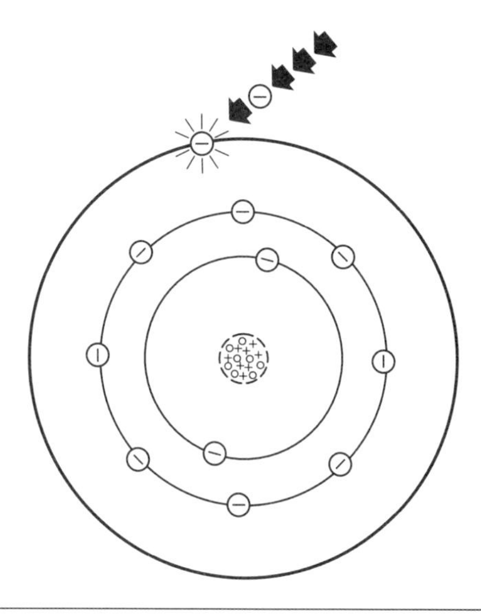

Figure 4-6 An electron being knocked out of orbit.

Silver is an excellent conductor of electricity because of its atomic makeup. There are five shells in a silver atom and the single valence electron in the outer shell orbits under a very high level of centrifugal force.

QUICK NOTE

The law of centrifugal force states that a spinning object will pull away from its center point. The faster an object spins, the greater the centrifugal force, and the greater the centrifugal force, the easier it is to knock the object out of orbit. Also, the greater the distance between an electron and the nucleus, the weaker the attraction between the two.

Silver is rarely used as a conductor in large quantities because of its high cost. However, it may be found on the contact points of some HVAC-R electrical components and in certain types of electronic components. An illustration of a silver atom is shown in Figure 4-7.

A more popular conductor of electricity is copper, shown in Figure 4-8. It is a less efficient conductor than silver, but is much more economical. A copper atom contains only four shells and the centrifugal force at which the valence electron spins is lower than that of silver.

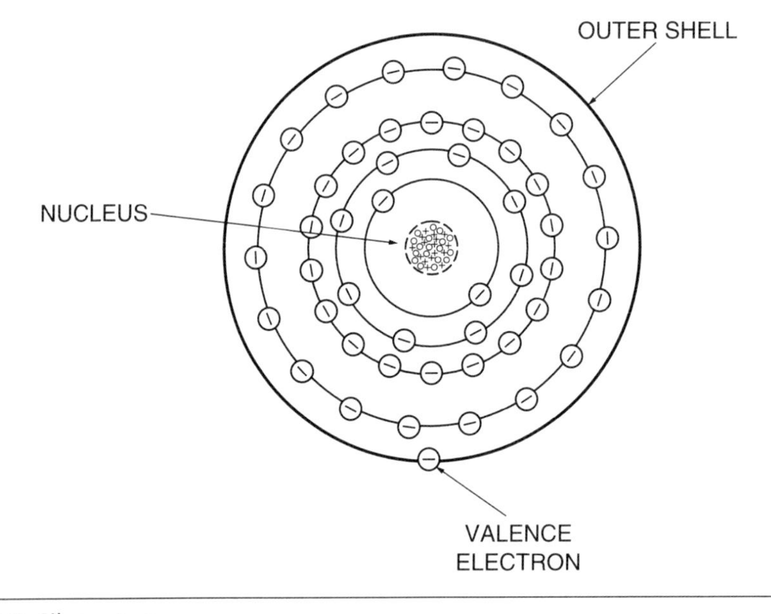

Figure 4-7 Silver atom.

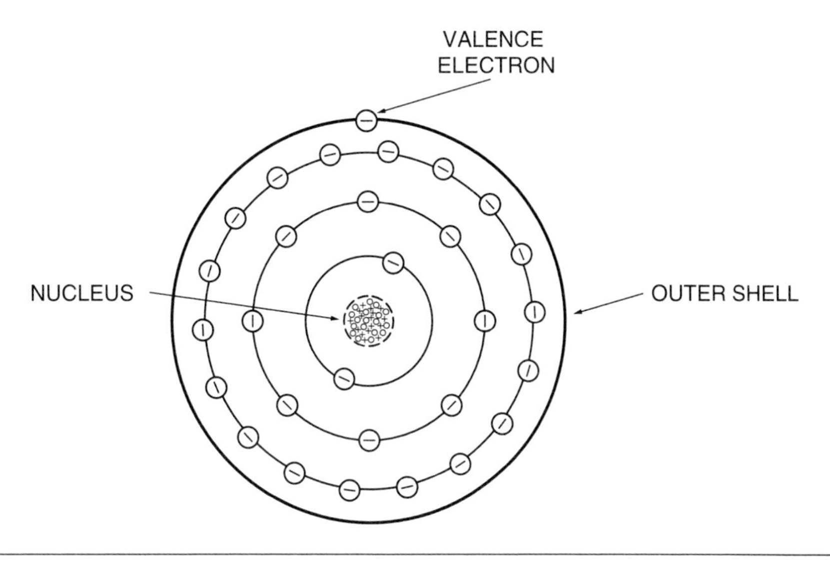

Figure 4-8 Copper atom.

Insulators

An insulator is a material with a high resistance to current flow. Its atomic structure is very different from that of a conductor. Instead of having one valence electron that is easily knocked out of orbit, the insulator has several valence electrons in its outer shell. The energy induced in the material is effectively dissipated among the valence electrons in an insulator and the current flow is stopped. An insulator atom is shown in Figure 4-9.

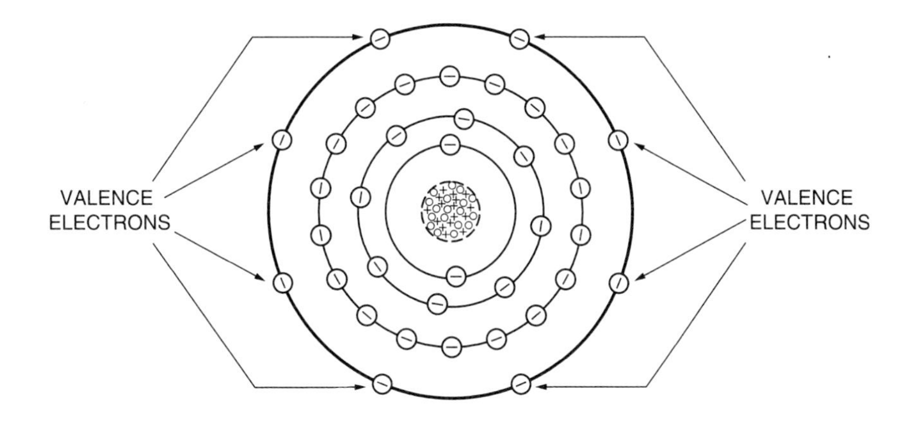

Figure 4-9 Energy induced in an insulator atom is effectively dissipated among the valence electrons.

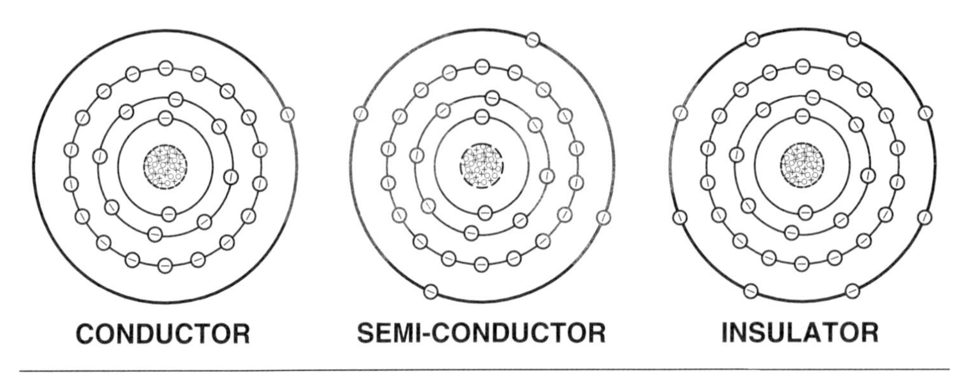

Figure 4-10 A comparison of conductor, semi-conductor, and insulator atoms.

While an insulator is effective in stopping current flow to a point, there is no perfect insulator and if the voltage is high enough, current flow cannot be stopped. All insulators will break down under the right conditions.

Semi-Conductors

As the name implies, a semi-conductor is neither an effective conductor nor an effective insulator. The number of valence electrons in a semi-conductor atom will lie somewhere between the single atom of a conductor and the higher number of electrons in the outer shell of an insulator atom. Semi-conductors are used in the manufacture of printed circuit (pc) boards.

Figure 4-10 shows a comparison between the atomic makeup of a conductor, semi-conductor, and insulator. A single valence electron is shown in the conductor, four valence electrons are shown in the semi-conductor, and the insulator atom has eight valence electrons.

CURRENT

Current is the rate at which electrons flow through a conductor. When measuring the amount of electrons that flow past a given point in a conductor, the term *ampere* is used. This term is derived from the French physicist Andre Ampere, who in 1820 was the first to measure the magnetic effect of electric current.

Any electrical component that performs useful work through the application of electrical energy has a given current draw when energized.

RESISTANCE

All conductors (other than superconductors) have some resistance to electron flow. Resistance is measured in *ohms*, a term that comes from the name of a German teacher by the name of Georg Ohm, who made discoveries in 1826 about the relationship between the electrical quantities of voltage, current, and resistance.

The fact that all conductors have some resistance to electron flow is the reason generating stations put the energy out at such a high level. Transmitting electricity at the level needed by the end user would result in no energy getting to the consumer.

OHM'S LAW

Ohm's Law illustrates the relationship between voltage, current, and resistance. When discussing Ohm's Law, letters are used to represent the electrical units. The letter E (for electromotive force) represents voltage; the letter I represents current (I for the intensity of the current in amps), and R represents resistance.

Ohm's Law applies to all systems using direct current (DC) and also to those using alternating current when working with purely resistive loads, such as heating elements. Inductive loads, such as those that are made up of coils of wire (motors, solenoids, etc.) cannot be considered workable under Ohm's Law because of the electromagnetic field set up by these types of components.

Ohm's Law can be used to determine any of the three values (E, I, or R) if the other two are known. For example:

$$E = I \times R$$

Multiplying the current (I) by the resistance (R) will give the applied voltage (E). To find current:

$$I = E \div R$$

Dividing the voltage (E) by the resistance (R) will give the current flow (I). To find resistance:

$$R = E \div I$$

Dividing the voltage (E) by the current (I) will give the resistance (R). An easy way to derive the equations for Ohm's Law is to use the memory wheel. With this method, you cover the factor you wish to find to display which equation is to be used. The memory wheel is shown in Figure 4-11.

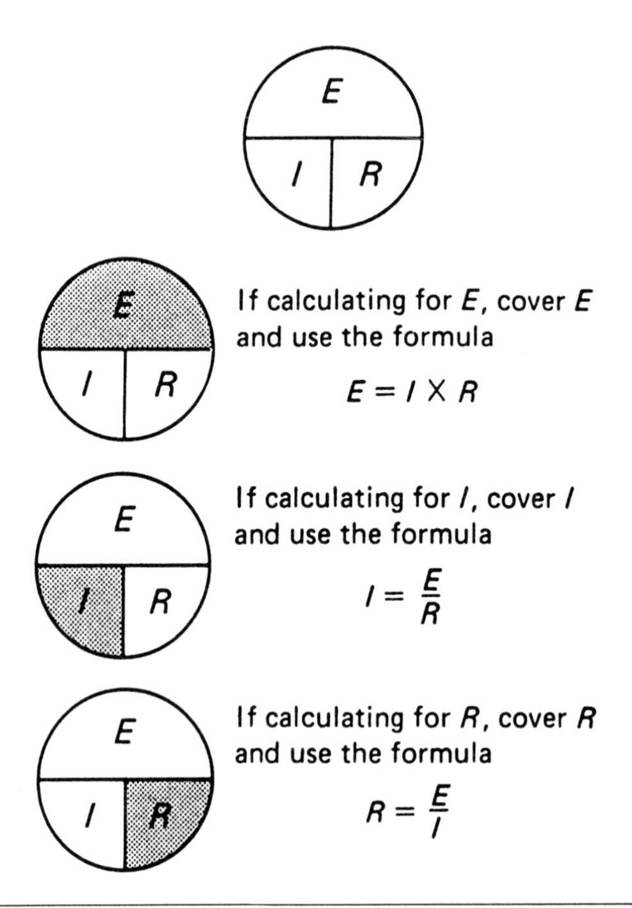

Figure 4-11 The Ohm's Law memory wheel can be used to solve electrical equations.

Ohm's Law can be used to determine if the resistance of a heating element in an electric furnace is within normal limits. For example, if the voltage applied to the element is 240 volts and the amperage draw of the element is 20 amps, the resistance of the element can be determined as follows:

$$R = E \div I$$
$$R = 240 \div 20$$
$$R = 12 \text{ ohms}$$

The calculated value can be compared against the measured value (taken with an ohmmeter) to determine if the resistance is correct.

ELECTRICAL POWER/WATTS

The watt is the measurement of true power. It is through the watt that the actual cost of operating a piece of equipment is determined. The electric company charges the consumer according to the number of kilowatts used. The term *kilowatt* (kW) refers to 1,000 watts.

The formula for calculating watts is as follows:

Volts x Amps = Watts

For example, an electric furnace operates on 240 volts and has a current draw of 20 amps. The wattage can be calculated as follows:

240 x 20 = 4,800 watts

The electric element in this furnace has an actual power consumption of 4,800 watts or 4.8 kW. With this information and knowledge of the rate charged per kilowatt/hour (kW/h) by the electric company, the cost of operating the furnace can be determined.

UNIT FOUR SUMMARY

To be an effective troubleshooter, you need to have a full understanding of electrical concepts, how electrical energy is generated, and how it is transported. Approximately 75 percent of the problems related to HVAC-R equipment are electrical in nature.

All HVAC-R equipment operates on alternating current. Alternating current is produced by cutting the lines of force of a magnetic field with a conductor. The source for alternating current is the AC generator. Rotating the generator at 3,600 RPM will achieve an energy delivery rate of 60 hertz.

Voltage is referred to as electromotive force and is the "electrical pressure" in a circuit. Conductors have few electrons in their outer shells, and readily permit the flow of electricity. Insulators have many electrons in their outer shells, and serve to halt the flow of electricity. A semi-conductor is neither a good conductor nor an effective insulator.

The resistance in an electrical circuit is measured in ohms, while current flow is measured in amperes. Ohm's Law is used to illustrate the relationship between voltage, resistance, and current.

The unit of measurement of true power in a circuit is the watt. The power consumed by a piece of equipment can be used to determine its operating cost.

Electrical Distribution Systems

To be an effective troubleshooter, you must be able to examine the "big picture" in order to determine the source of a problem. For example, if you have an air conditioning unit that is not working properly, the problem may rest with a failed component within the unit itself, or it may be in the air system, piping, construction of the home, or electrical system. In this unit, we will concentrate on the electrical distribution system. Improper installation or modification of an electrical distribution system can cause inadequate operation or even equipment failure.

As an HVAC-R technician, you must be able to rule out supply voltage problems as the cause of equipment failure or inadequate performance. In the event that the electrical supply system is at fault, you have to make your customer understand that although the equipment itself is not operating, the problem lies with the electrical supply and an electrician must be called in to correct it.

SINGLE-PHASE AND THREE-PHASE SYSTEMS

Alternating current is delivered to the end user in several voltage and current configurations. Residential buildings are usually supplied with systems that are rated at 240 volts and described as being single phase, 60 hertz. In a single-phase system, one set of windings is used to cut the lines of force of a magnetic field.

Commercial buildings may also use single-phase systems, but are usually three-phase systems that supply 240 volts, 460 volts, and 208 volts. In a three-phase system, three sets of windings are used to cut the lines of force of a magnetic field.

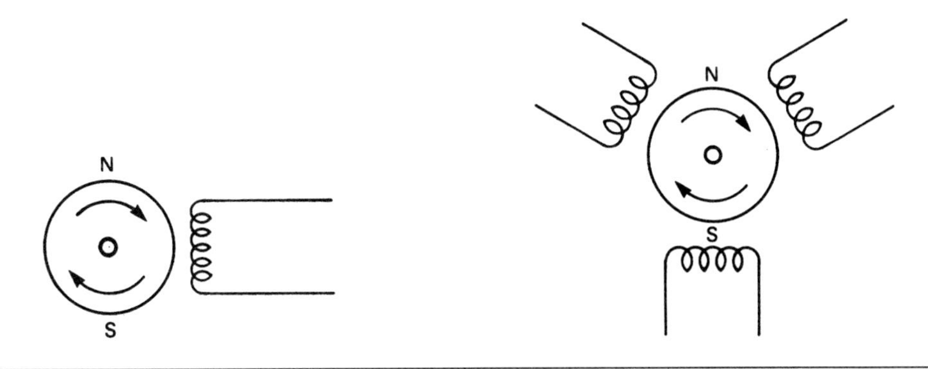

Figure 5-1 The winding arrangements of single-phase and three-phase current generators.

QUICK NOTE

A *phase* is described as the number of currents alternating within a circuit. Single-phase and three-phase alternating current generators are shown in Figure 5-1.

SINGLE-PHASE POWER

In residential areas, power is delivered from the secondary winding of a transformer that is located either on a pole or on a pad. Figure 5-2 shows the power supply to a residence in an overhead application, and Figure 5-3 shows underground service from a pad-mounted transformer.

Figure 5-2 The power supply to a residence in an overhead application.

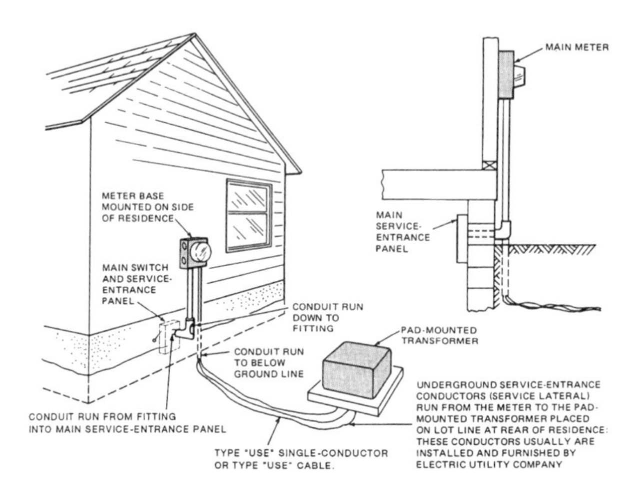

Figure 5-3 The power supply to a residence in an underground application.

After the wiring is routed to the house, current is channeled through a meter and then on to what is known as the *main disconnect panel* or *main service entrance panel*. Within the panel are the bus bars, and attached to the bus bars are the circuit breakers that control the various circuits within the residence.

QUICK NOTE

Instead of circuit breakers, some older buildings may use cartridge type fuses as a main disconnect, and screw-in type fuses (known as *plug fuses*) to control the individual circuits within the residence.

Figures 5-4, 5-5, and 5-6 illustrate the residential service panel and provide a general idea of how circuit breakers are placed in the distribution box.

Figure 5-4 shows the "hot" bus bars located within the distribution panel. The two legs of power are connected to the bus bars, and this is where the 240-volt power supply system begins. Also within the distribution panel is the "neutral" bus bar. In a residential system, connecting one wire to one of the hot legs and one wire to the neutral bar will render the 120 volts that supply power to lighting, small kitchen appliances, refrigerators, and other items that are plugged into standard wall outlets.

Figure 5-5 shows the same panel with the addition of circuit breakers. A single pole breaker locks onto only one bus bar. It supplies one hot leg of power to a 120-volt wall outlet or light. To provide a complete 120-volt circuit, another wire to the 120-volt equipment receptacle or lighting is connected to the neutral bus bar.

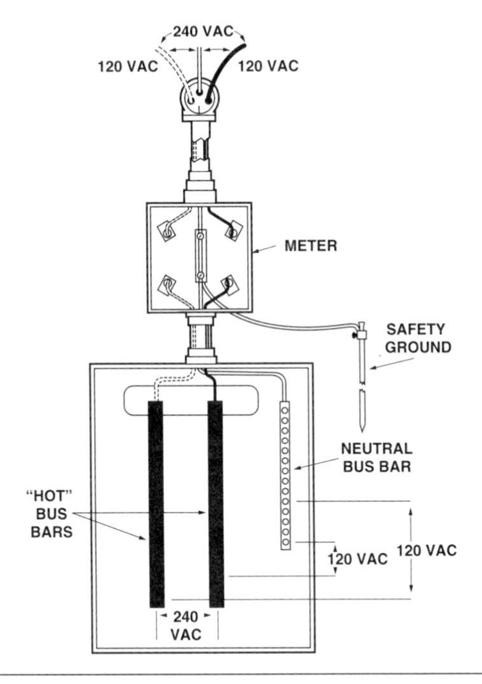

Figure 5-4 The "hot" bus bars in an electrical distribution panel.

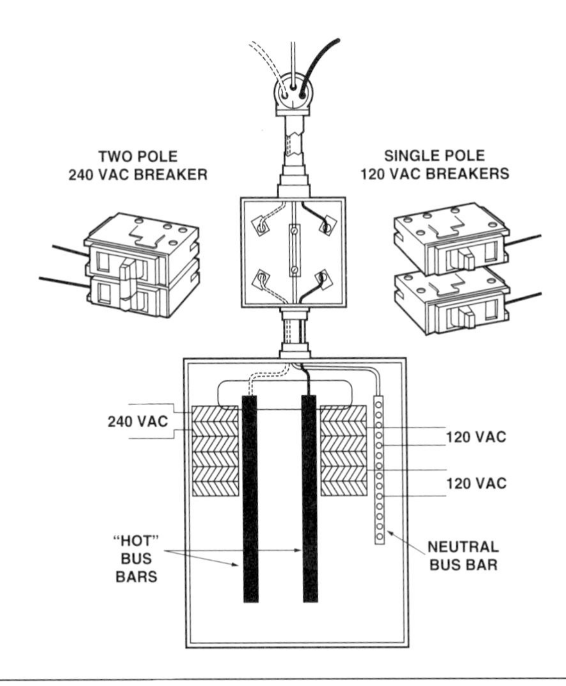

Figure 5-5 The circuit breakers in an electrical distribution panel attach to the bus bars.

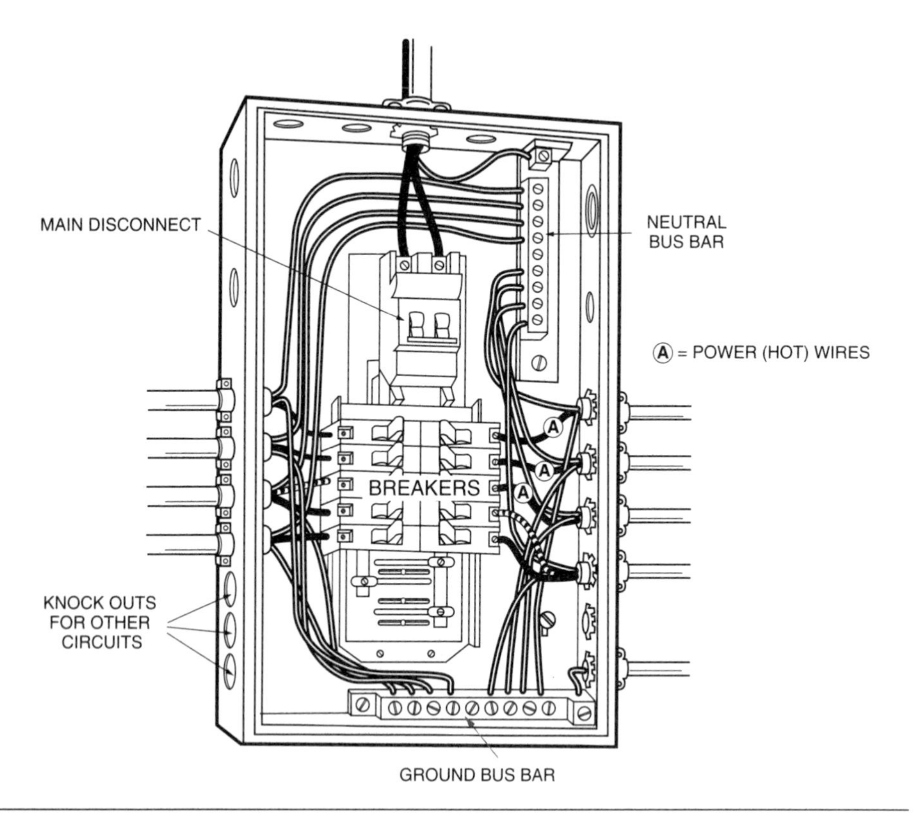

Figure 5-6 A detailed drawing of a residential distribution panel.

A double pole breaker is one that locks onto both bus bars. It provides a 240-volt circuit to equipment such as an air conditioner, electric furnace, electric water heater, electric range, or electric dryer.

Figure 5-6 shows the residential panel in more detail. The circuit breakers have been attached to the bus bars and the wiring that provides power to the circuits throughout the house has been added. In our illustration, a main disconnect breaker is shown. Not all residential panels are equipped with a main disconnect.

WARNING

Always exercise extreme caution when removing the cover panel from a main panel to inspect the circuit breaker system. Remember, turning the circuit breaker to the OFF position does not cut the power to the main wiring that connects to the bus bars. Even if all the breakers are off, the main wiring to the bus bars is still hot!

THREE-PHASE POWER

Three-phase systems are found in commercial and industrial applications. After the initial cost of panel and wiring installation, these systems save costs in operating motors and other equipment. Three-phase systems get their power from the secondary of a transformer that is designed in either a delta or a wye configuration.

Delta Systems

Three-phase systems have three hot legs instead of two. Figure 5-7 shows a schematic of the secondary windings of the delta transformer. This system is known as a *delta arrangement* because the configuration of the windings resemble the Greek letter delta (Δ).

With this type of electrical system, a reading of 240 volts will be shown between any two of the hot legs. Also, placing the leads of a meter between L2 and neutral or L3 and neutral will yield a reading of 120 volts just as it would in any residential system. The difference with the three-phase delta system is the reading between L1 and neutral. L1 is what is known as the *high leg* or *wild leg*. A reading between L1 and neutral will lie somewhere between 180 and 208 volts.

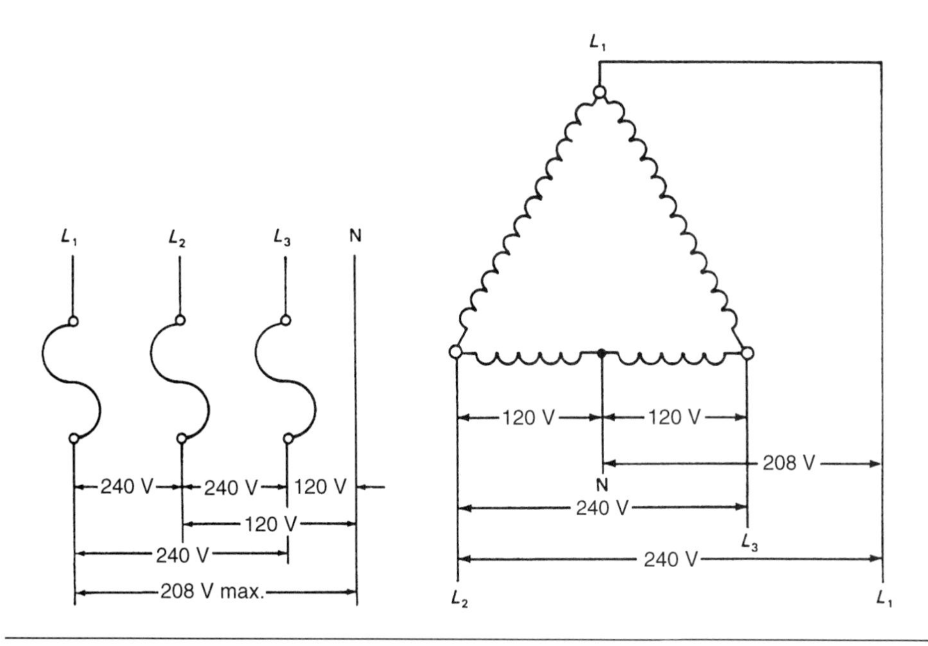

Figure 5-7 Schematic and pictorial for a delta transformer system.

Connecting any 120-volt piece of equipment between L1 and neutral will result in damage to the equipment. Technicians often used a "cheater cord," which is a standard 120-volt female plug wired to alligator clips, to temporarily connect to a rooftop disconnect box and operate drills or vacuum pumps. Connecting a cheater cord to L1 and neutral on a delta system and then plugging in a vacuum pump or other 120-volt equipment would be an expensive lesson on the "wild leg" of an electrical system.

Wye Systems

The other type of three-phase system used in commercial buildings is the *wye system*. It resembles the letter Y. This system differs from the delta system in that it registers 208 volts between any two hot legs. The wye system, such as the one shown in Figure 5-8, has no high leg.

Not all three-phase systems supply 240 or 208 volts. Higher voltage systems, such as 277/480 volts, are used in some commercial applications. A higher voltage system may save the end user money due to the fact that the wiring to equipment can be a smaller gauge. (The smaller the gauge, the more inexpensive the wiring.)

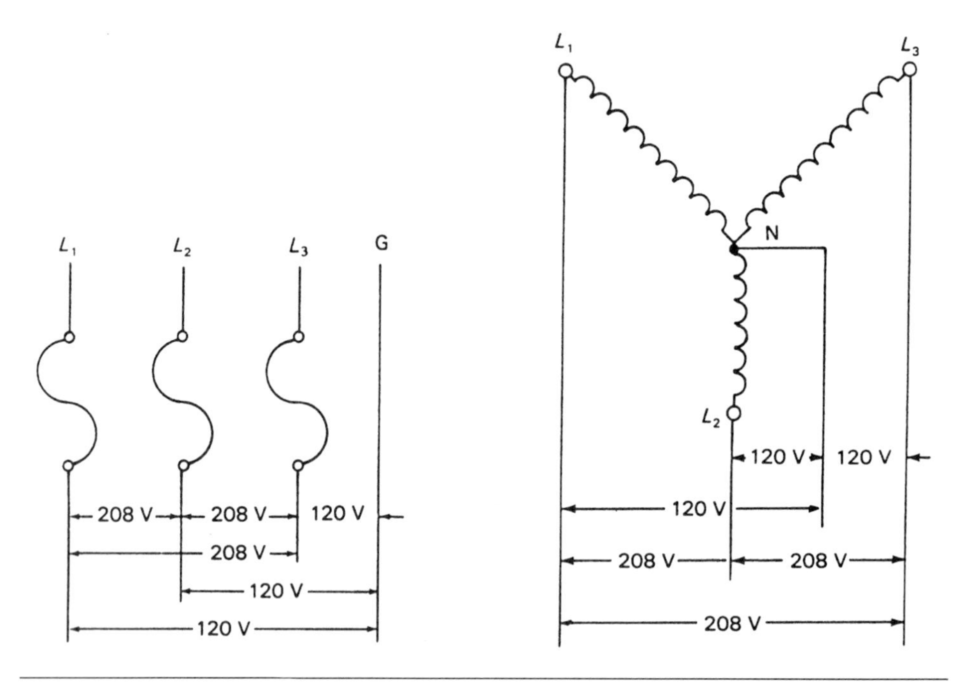

Figure 5-8 Schematic and pictorial for a wye transformer system.

UNIT FIVE SUMMARY

A troubleshooter must be able to determine whether the problem with a piece of equipment lies within the equipment itself or in one of the supporting systems, such as the electrical distribution system.

Residential electrical systems are supplied with single-phase power. Wiring from the secondary of the supply transformer is connected to bus bars in a main disconnect panel, and circuit breakers connected to the bus bars supply power to the individual circuits. Some older homes may use cartridge fuses and plug fuses instead of circuit breakers.

Commercial and industrial buildings use three-phase power. Two types of transformer windings found in three-phase systems are the delta and wye configurations. The delta system is arranged in a triangular formation like the Greek letter delta. It registers 240 volts between any two hot legs and has a high leg that can damage 120-volt equipment. The wye system is arranged like the letter Y. It registers 208 volts between any two hot legs.

In some cases, higher voltage systems (such as 277/480 volts) may be found in commercial buildings.

HVAC-R Electrical Components

Now that you have an understanding of refrigeration and electrical fundamentals, we will proceed with a discussion of some of the components that make up an HVAC-R system. Within any piece of equipment, a component either uses electricity to perform work and is defined as a *load*, or it controls current flow and is defined as a *switch*. When troubleshooting a system, you must be able to identify failures in both types of components. For example, an inoperative motor may be caused by a failure of the motor itself, or it may be caused by the relay that controls the motor power circuit.

ELECTRIC MOTORS

Electric motors perform a wide range of functions within HVAC-R equipment, including air handling, pumping water, and operating refrigeration system compressors. To understand the different types of motors used and their many applications, you first need to understand the basic operation of an elementary electric motor. As discussed in Unit 4, when current flows through the conductor in an electric motor, a magnetic field is set up around the conductor.

You will recall that the law of electric charges states that opposite poles attract and like poles repel. This is the principle behind the operation of an electric motor. Figure 6-1 shows an elementary electric motor and its components: the *stator*, which is the stationary part of the motor and is actually an electromagnet, and the *rotor*, which is the part of the motor that rotates.

As you can see, two factors determine the rotation of the rotor portion of the motor:

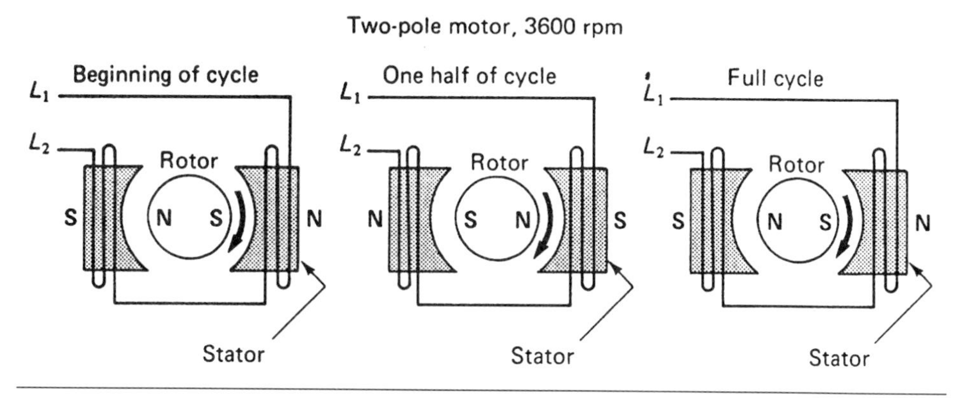

Figure 6-1 The rotor and stator of an elementary electric motor.

1. The rotor is what is known as a *permanent magnet.*
2. The polarity of the electromagnet that makes up the stator section of the motor changes. Remember, in order to generate 60 Hz AC, the lines of force of a magnetic field are cut 60 times per second. In other words, the alternating current reverses its direction 120 times per second. This causes the permanent magnet of the rotor to be attracted and then repelled each time the reversal occurs.

Split-Phase Motors

Simply defined, a *split-phase motor* is a motor that has a run winding and a start winding. You may wonder why there is a need for a separate start winding. Consider this: when the rotor of the motor assembly is at rest, the magnetic field set up in the motor windings must be strong enough to initiate the rotating action. Once the rotation of the rotor has begun, the amount of energy needed to keep it moving is not as great because the rotor has momentum. So, while a high level of energy is necessary to start the motor, a smaller amount of energy is required to keep it running.

The start winding of a motor is constructed of wire that is longer and thinner than that of the run winding. This creates a higher resistance, which results in the stronger magnetic field necessary for the starting process. Because the amperage draw of the start winding is larger than the normal current draw, it is designed to stay energized for only a short period of time.

Figure 6-2 When a centrifugal switch is used to break the circuit to the start winding of a motor, it is located at the rear of the motor. (By Bill Johnson)

The starting/running process is accomplished by energizing both the start and run windings at the instant of start, then deenergizing the start winding after it has done its job of getting the rotor started. This can be accomplished in a variety of ways, depending on the design of the motor and the start components used. A *centrifugal* switch is a common method of breaking the circuit to the start winding. Centrifugal switches can be found in motors that are used in evaporative coolers or in air handling systems in furnaces and air conditioners. Figure 6-2 shows a centrifugal switch, which is located at the rear of the motor.

The centrifugal switch opens when the motor reaches about 75 percent of its running speed. If the centrifugal switch does not break the circuit to the start winding, the motor runs for a short period of time at a speed somewhat slower than its designed running speed, then kicks off on overload. A protective device built into the motor senses the extended period of high current draw, and breaks the power circuit to the motor entirely. Some protective devices automatically reset and allow the motor to attempt to restart after a certain period of time. In some cases, you may find a motor that has a manual reset button.

Whether it is equipped with a manual or automatic reset, the wiring for the centrifugal switch system is the same. Figure 6-3 shows a simplified schematic wiring diagram for a motor that uses a centrifugal switch.

When tracing this diagram, you can see that when the switch is in the closed position, the 125-volt power source is being applied to both windings.

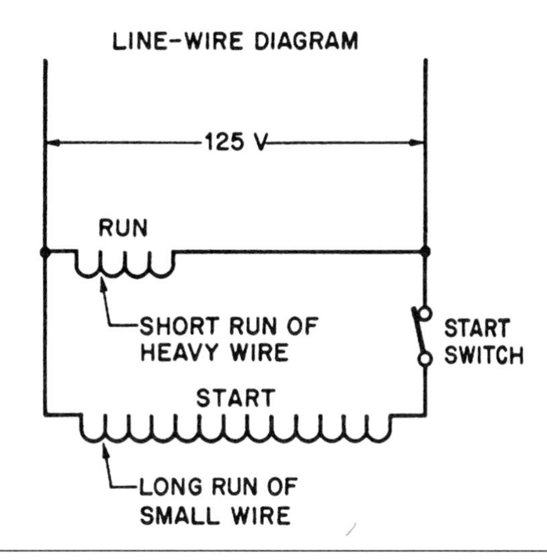

Figure 6-3 A wiring diagram of a motor start winding, run winding, and start switch.

This is what is known as a *parallel circuit*; that is, an equal amount of power is applied to two separate loads. However, when the centrifugal force produced by the rotor reaching a high speed forces the switch to open, the circuit to the start winding is broken, and the motor continues to operate on the run winding alone.

Figure 6-4 shows a pictorial diagram of the same split-phase motor circuit and centrifugal switch system.

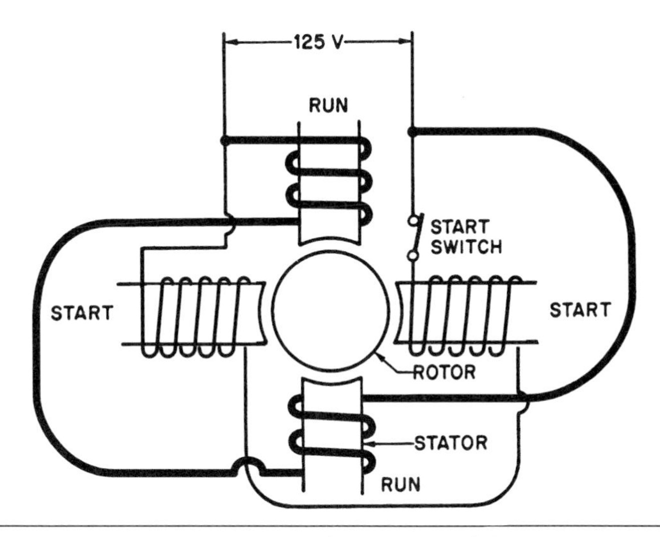

Figure 6-4 A pictorial diagram showing the routing of the wiring in a motor.

This diagram shows the actual routing of the wiring in the motor instead of just showing the circuit electrically. It is more difficult to read than a schematic, but the circuit to both windings can be traced if you take a step-by-step approach.

Starting at the left of the 125V power source, follow the thin wire as it is routed first around the start winding section of the stator on the left, then as it is routed directly to and around the start winding section of the stator on the right. You will notice that this wire is routed through the start switch (centrifugal switch) before returning to the power source. This is the complete path of the start winding circuit.

To trace the run winding circuit, start again at the 125V power source on the left, but this time follow the thinner wire only to the point at which there is a connection to the run winding circuit, which is shown as a heavier line. Trace the heavy wire as it is routed directly to the stator section at the top of the motor assembly, then is routed directly to and around the stator section located at the bottom of the motor. The run winding circuit is complete when the wire is routed back to the 125V power source. Note that there are fewer turns of wire around the stator in the run winding than there are in the start winding.

Figure 6-5 gives you another view of the start and run windings of a motor. A small section of the stator is shown, and you can see that there are more turns of wire in the start winding than there are in the run winding.

The split-phase motor is often referred to as a *squirrel cage motor* because of the cage-like appearance of the rotor section of the motor. The rotor and its position in the stator assembly are shown in Figure 6-6.

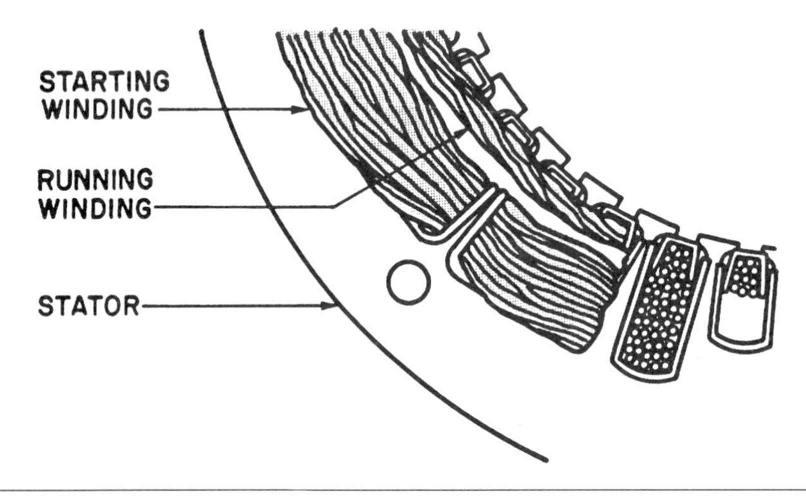

Figure 6-5 How the start and run windings are placed in a stator.

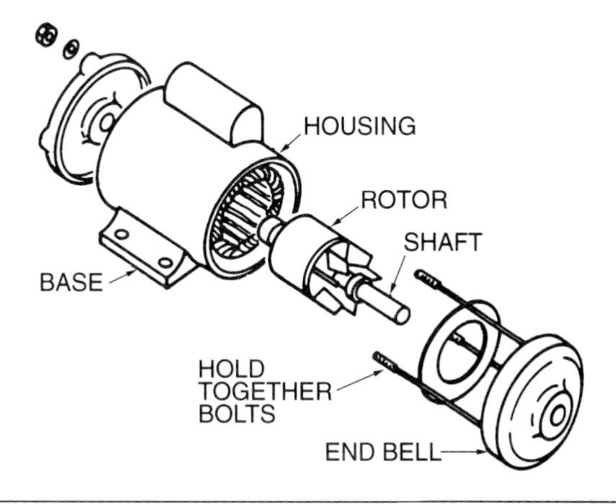

Figure 6-6 The individual components of an electric motor.

The squirrel cage rotor itself (shown in Figure 6-7), is not a permanent magnet such as that used in an elementary electric motor. In a squirrel cage motor, the rotor is made up of either copper or aluminum bars spaced evenly around the iron core that makes up the body of the rotor assembly.

While the rotor itself is not a permanent magnet, it becomes a magnet through the process of induction. The primary magnetic field is the stator of the motor, and the rotation of the rotor in this magnetic field induces a current in the iron core. When the magnetic field is induced in the rotor, the magnetic poles of the rotor do not alternate, so it does, in effect, act like a permanent magnet. The poles of the electromagnet in the stator do alternate from north to south as we have previously described, and as a result, the motor operates in the same manner as an elementary motor.

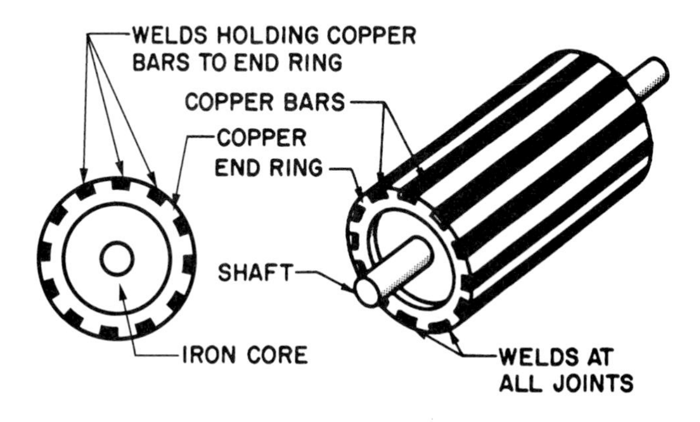

Figure 6-7 The rotor of a squirrel cage motor.

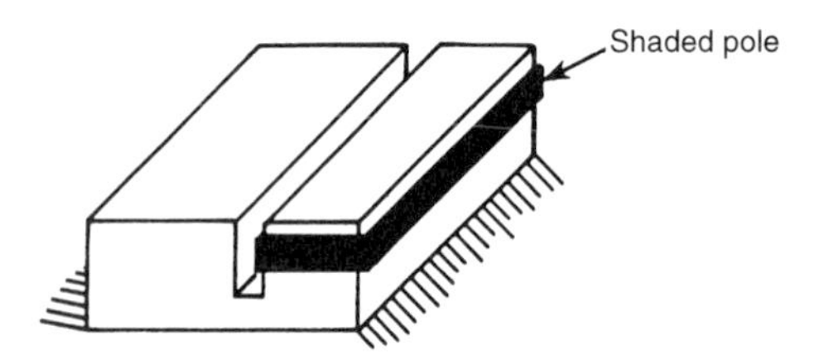

Figure 6-8 Instead of a separate winding, the shaded-pole motor uses a solid copper band that has been fitted into a groove cut into the stator.

Shaded-Pole Motors

Another type of motor you may find in a furnace air handling system or as a small condenser fan is the shaded-pole motor. This motor operates differently from the split-phase motor in that it does not use a separate start winding. There are stator windings, but they serve to induce an out-of-phase magnetic field into a solid copper band or very heavy wire that has been fitted into a groove cut into the stator. The current induced in the band by the stator winding is sufficient to provide enough torque to get the rotor moving. An illustration of the method of construction of a shaded-pole motor is shown in Figure 6-8.

The shaded-pole motor has a much lower starting torque than the split-phase motor and can stall easily.

In some cases, the shaded-pole motor is designed as a multi-speed motor.

CAPACITORS

A simple definition of a *capacitor* is that it is an electrical storage device. It consists of two metal plates separated by a dielectric (nonconductive) material, usually an oil. The special construction of a capacitor allows it to store an electrostatic charge in the metal plates, and then release that charge to provide a boost of power. Two common types of capacitors are the start capacitor and the run capacitor.

Permanent Split Capacitor (PSC) Motors

The PSC motor is one of the most popular motors in the HVAC-R industry. Used in air handling systems, condenser fans, and hermetic compressors, the windings in a PSC motor are much closer in diameter and length of wire than those in a standard split-phase motor. With a PSC motor, no switching device is used to break the circuit to the start winding.

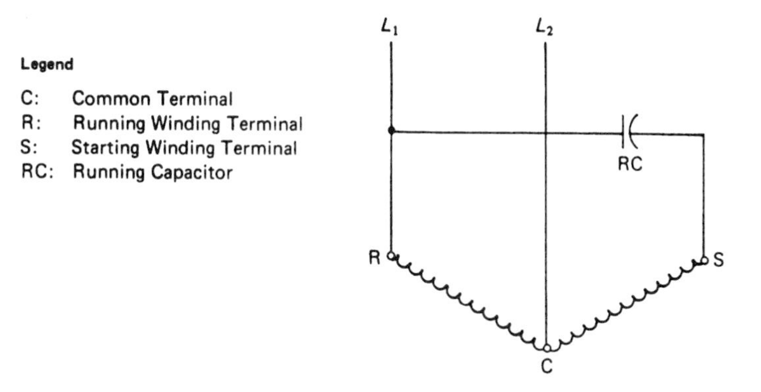

Figure 6-9 A schematic representation of a PSC (permanent split capacitor) motor.

Instead, the start winding is always energized, just as the run windings are, but a run capacitor wired in series with the start winding is used. An illustration of a PSC motor is shown in Figure 6-9, and a run capacitor is shown in Figure 6-10.

In a PSC motor, the run capacitor is used to increase the operating efficiency of the motor. As you can see from Figure 6-9, the run capacitor is designed to be in the circuit all the time. Power leg L2 is wired directly to the common terminal on the motor to provide one side of the circuit for both windings in the motor. Power leg L1 provides the other side of the line by being wired directly to the run winding, while at the same time being wired to the start winding through the run capacitor.

Figure 6-10 A run capacitor is used to improve a motor's operating efficiency.

Run capacitors are rated in a unit of measurement known as the *micro-farad*. The run capacitor is oil-filled for the purpose of dissipating heat.

Sizing Run Capacitors – If you find it necessary to replace a run capacitor, be sure to use a replacement that is rated within industry guidelines. These include the following:

1. The voltage of the replacement capacitor must be equal to or greater than the capacitor being replaced.
2. The strength (microfarad rating) of a replacement run capacitor must be within plus or minus 10% of the original capacitor.

Failure to follow these guidelines could result in damage to the motor.

Figure 6-11 shows a pictorial representation of a PSC circuit for a hermetic compressor. As in the schematic, one leg of power is run directly to the common terminal of the motor assembly (L1), while L2 is used to apply power directly to both the run capacitor and the run winding. The start winding is powered by L2 as well, but its circuit is wired through the run capacitor. A PSC motor is considered to have a moderate starting torque.

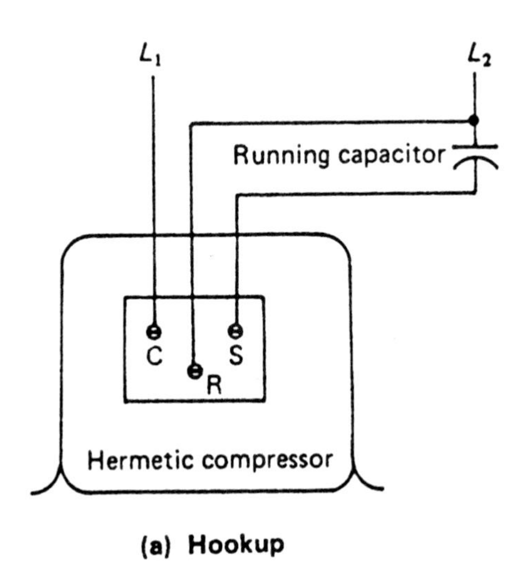

Legend

C: Common Terminal
R: Running Winding Terminal
S: Starting Winding Terminal
RC: Running Capacitor

Figure 6-11 A pictorial illustration of a PSC motor.

Figure 6-12 A start capacitor is used to provide a momentary boost.

Capacitor Start (CS) Motors

In applications where a motor needs to start under a load, a start capacitor is used. It is sometimes referred to as an *electrolytic capacitor*. Like the run capacitor, the start capacitor is wired in series with the start winding, but does not stay in the circuit through the run cycle of the motor. The job of the start capacitor is to provide a momentary boost to the start winding and increase the starting torque of the motor.

A start capacitor, such as the one shown in Figure 6-12, differs from a run capacitor in that it is smaller and contains less dielectric material. It is not designed to remain in the circuit for a long period of time.

Miswiring a start capacitor and leaving it in the circuit after starting the motor will result in damage to the capacitor. A simplified capacitor start schematic wiring diagram is shown in Figure 6-13.

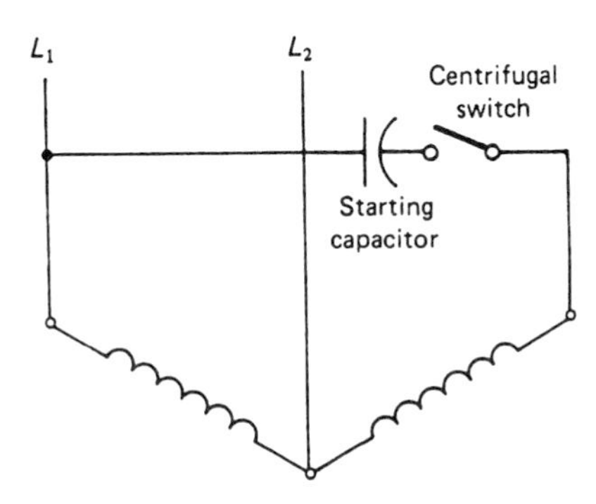

Figure 6-13 Schematic diagram of a capacitor start motor with a centrifugal switch.

As you can see in the diagram, the switch that breaks the circuit to the start winding is "downstream" from the start capacitor. Placing the switch after the start capacitor effectively removes it from the start winding circuit.

START-ASSIST RELAYS

The two types of start-assist relays commonly used on compressor motors are the potential relay and the current relay.

Potential Relays

Potential relays can be found in the *hard start kits* that are sometimes used with compressor motors to provide extra starting torque. This relay derives its name from the fact that it operates on the potential (something referred to as *back EMF*) created by the start winding of a motor when it is operating. Back EMF is short for *back electromotive force*, another name for voltage. A potential relay is shown in Figure 6-14.

From an electrical standpoint, there are three terminals you have to understand when troubleshooting a potential relay. They are terminals 1, 2, and 5. Potential relays may have other terminals, such as 3, 4, and 6, but they are only used as tie points for convenience when wiring the electrical circuit. Terminals 1, 2, and 5 are the important ones.

Refer to Figure 6-15. This simplified illustration of a potential relay shows that there is a coil between terminals 2 and 5. It also shows that there is a switching assembly between terminals 1 and 2. A wire is shown connected from terminal 2 to the compressor start winding, and a second wire is shown connected from the common terminal of the compressor to terminal 5 on the potential relay. When the compressor is running, the back EMF created by the start winding is what energizes the coil section of the potential relay.

Figure 6-14 A potential relay is one component of a hard start kit.

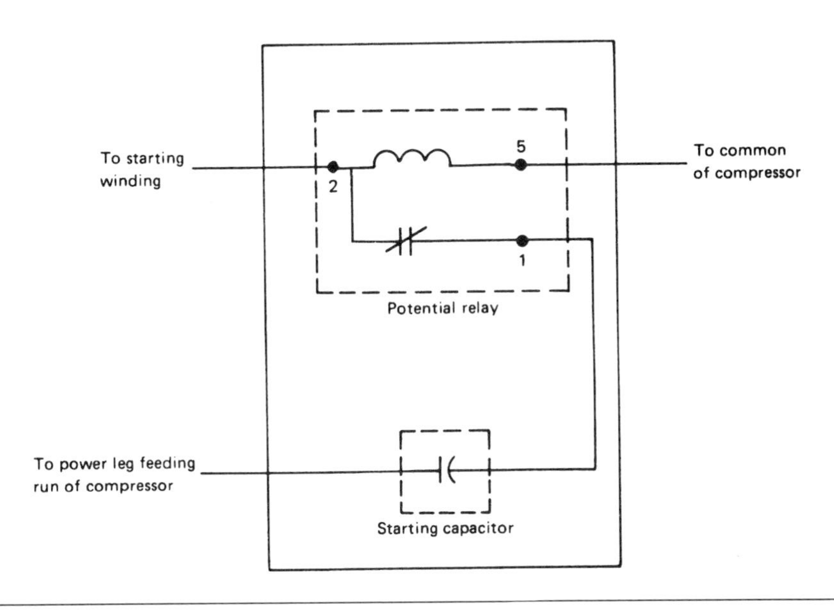

Figure 6-15 A schematic diagram of a potential relay.

The maximum back EMF is created when the motor reaches about 75% of its normal running speed. When the back EMF applied is sufficient, the magnetic field of the coil in the potential relay is strong enough to cause the switch inside the relay to open, breaking the circuit from the start winding to the power leg that feeds both the run and start windings of the compressor. A simplified diagram of a potential relay is shown in Figure 6-16.

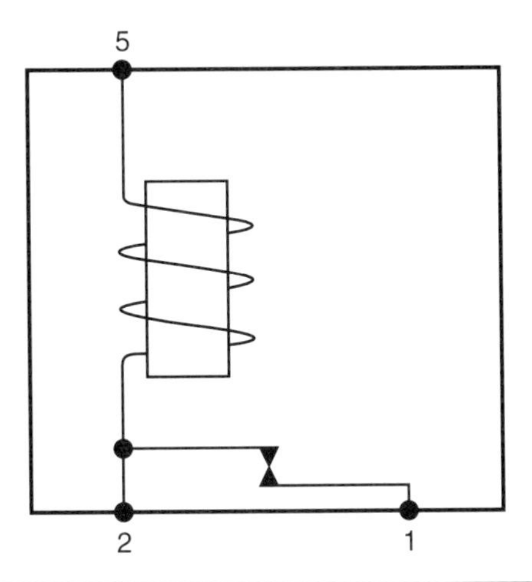

Figure 6-16 A simplified diagram of a potential relay.

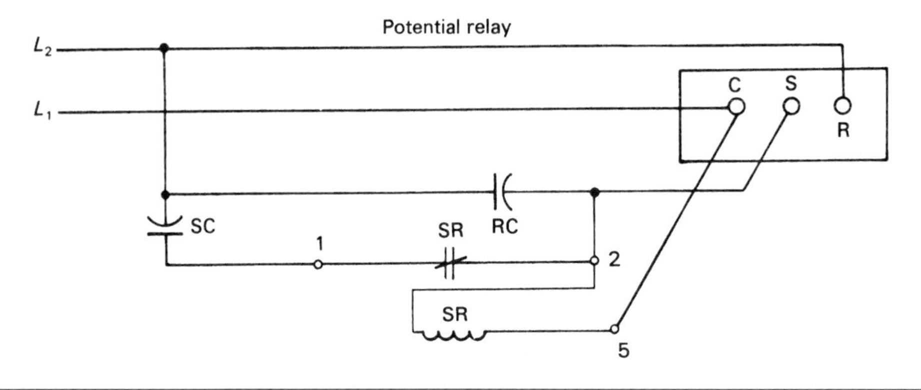

Figure 6-17 A schematic diagram showing a capacitor start/capacitor run motor and a potential relay.

A schematic diagram of a potential relay in a compressor circuit is shown in Figure 6-17. Keep in mind that although the start relay switch (SR) and the start relay coil (also SR) are shown separately, they are interconnected in the circuit in such a way as to allow the switch to be operated by the coil.

In this diagram, L1 is shown wired directly to the common terminal of the compressor and also directly to terminal 5 on the potential relay. Remember that this is shown schematically and it appears that the wire is connected first to the common compressor terminal, then is routed to terminal 5. In a real application, terminal 5 on the potential relay may have a place for the connection of two wires. This would then be the tie point for the wiring that connects both the common terminal and terminal 5.

The wire identified as L2 on the schematic is shown connected in several places. First, you can trace the wire that is run directly from L2 to the run terminal on the compressor. The same wire also provides power to the run capacitor (RC) and the start capacitor (SC). The circuit through the run capacitor runs directly to the start winding, but the circuit through the start capacitor must run through the SR switch before getting to the start winding.

At the instant of start, there is a complete circuit from L1 to common, from L2 to run, and from L2 to start through the run capacitor, start capacitor, and SR switch. When the back EMF generated by the start winding supplies the *pickup voltage*, the coil of the relay is energized and the SR switch breaks its circuit to the start winding. The RC circuit to the start winding remains after the SR switch opens.

The two things to keep in mind about pickup voltage are:

1. The pickup voltage of a potential relay will be higher than the voltage applied to the compressor in the first place.

2. The pickup voltage supplied by compressors varies. A potential relay that matches the compressor pickup voltage must be used.

In most cases, you can determine which relay to use on a compressor by using compressor manufacturer's information that lists the proper relay, start capacitor, and run capacitor.

Relay manufacturers also list information regarding the pickup voltages of the relays they make. One method of determining the pickup voltage of a particular compressor is shown in Figure 6-18.

As you can see, voltmeter 1 is showing the applied voltage to the compressor at 125 volts, and the probes of the meter are attached to the common and run terminals. Voltmeter 2 is shown measuring the pickup voltage from the start winding and is connected to the common terminal and the start winding of the compressor.

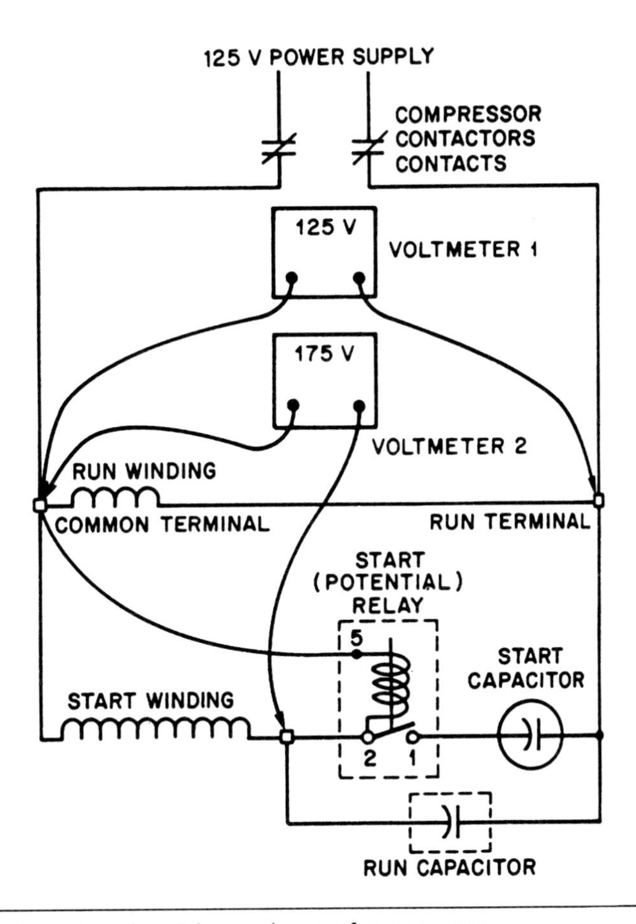

Figure 6-18 Measuring the pickup voltage of a compressor.

Sizing Potential Relays – In the event that you need to size a potential relay, follow this procedure:

STEP ONE: Start the compressor manually with a test cord.

STEP TWO: Check the voltage between common and start with the compressor running at full speed.

STEP THREE: Multiply the voltage reading by 0.75 to calculate the pickup voltage of the relay.

STEP FOUR: Check the specifications sheet of the potential relay to be sure that its operating parameters are within the voltage you calculate.

Current Relays

Another type of start relay that is used on refrigeration compressors is the current relay (Figure 6-19). It is commonly used in fractional horsepower compressors in refrigerators and some bar and restaurant equipment. The current relay is also referred to as a *magnetic relay* or *amperage relay*. This type of relay usually plugs onto the terminals of the compressor.

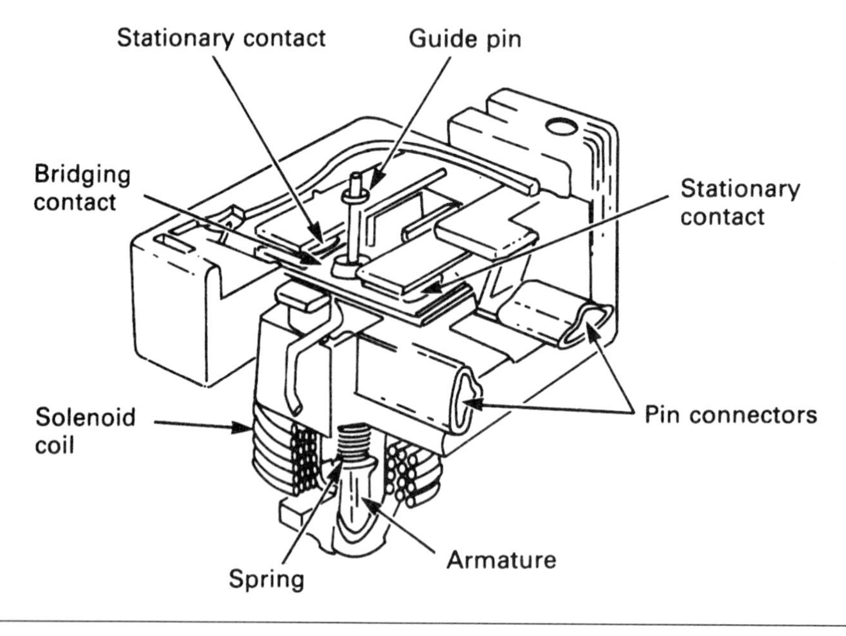

Figure 6-19 A cutaway of a current relay, also referred to as an amperage relay. (Adapted with permission from Texas Instruments, Inc., Attelboro MA)

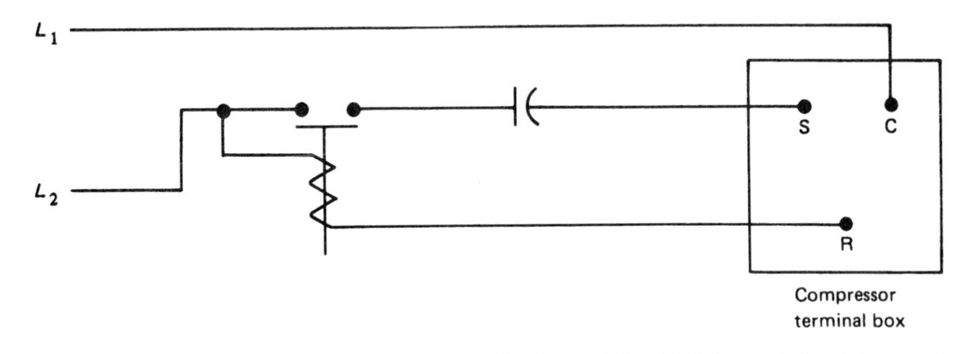

Figure 6-20 A schematic diagram showing a current relay when the coil is deenergized.

A current relay differs from a potential relay in that the coil of the relay is wired in series with the run winding of the compressor. Instead of working off the voltage supplied by the start winding, the current relay makes and breaks the circuit to the start winding according to the amperage draw of the run winding. A schematic diagram of a current relay is shown in Figure 6-20.

The sequence of operation of a current relay is as follows:

1. With no power applied to the compressor, the contacts of the relay are open.
2. At the instant of start, the current draw of the run winding is very high. As a result, the magnetic field of the coil in the relay is also high, causing the solenoid-like mechanism to move. This makes (closes) the contacts of the relay in the circuit to the start winding.
3. After the start winding is energized and the compressor motor reaches 75 percent of its running speed, the current draw of the run winding drops, weakening the magnetic field of the coil. The result is that the mechanism that provided the closure between the two points now drops, breaking the circuit to the start winding.

Sizing Current Relays – Like a potential relay, the current relay must also be matched to the compressor. For example, using a relay for a ⅓-horsepower compressor on a smaller compressor would not allow proper operation of the compressor because the current draw of the run winding would not be high enough to set up the magnetic field on the relay coil and close the contacts to the start winding.

Similarly, using a relay that is too small could also result in a problem. The coil of a relay designed for a smaller compressor may keep the contacts to the start winding closed because the current draw of the run winding is too high.

MOTOR OVERLOAD PROTECTORS

Motors in HVAC-R equipment can become overloaded for a variety of reasons. For example, a belt could be adjusted too tightly, the voltage being supplied to a motor could be too low due to a poor connection in a disconnect box or breaker, or a refrigeration system may be overcharged. Whatever the cause, in order to protect a motor from overheating and burnout, protective devices must be used.

Thermal Overloads

The two types of thermal overload devices commonly used in HVAC-R equipment are the internal overload and the external overload.

Internal Overload Devices – In many motors, such as those in air conditioning compressors or air handling systems, an internal thermal overload embedded in the motor windings is used. An internal overload protective device is shown in Figure 6-21.

Figure 6-21 An internal overload is fitted into its own hermetic container when used in a refrigeration compressor. This prevents the contamination of refrigerant by the electric arc. (Courtesy Tecumseh Products Company)

Figure 6-22 An externally-mounted overload protector.

TROUBLESHOOTING HINT

When troubleshooting electric motors with an internal overload protection device, wait for the motor to cool off before using an ohmmeter to test the windings of the compressor. Testing when the compressor is hot will show an incorrect reading of an "open winding," when in fact the overload is operating properly.

External Overload Devices – On motors built without internal protection, such as fractional horsepower compressors used in refrigerators and freezers, an external overload protector is used. In most cases, these are wired in series with the common terminal of the compressor. An external overload protective device is shown in Figure 6-22.

TRANSFORMERS

Transformers found in HVAC-R equipment are commonly referred to as *step-down transformers*. A step-down transformer is a device that accepts a given voltage at its primary winding and delivers a lower voltage from its secondary winding. This reduction in voltage is accomplished by wrapping an iron core with a given number of windings on the primary side, then wrapping the same iron core with a fewer number of windings on the secondary side. Transformers are used in HVAC-R circuits to convert high voltage power into the lower voltages needed for the control circuit.

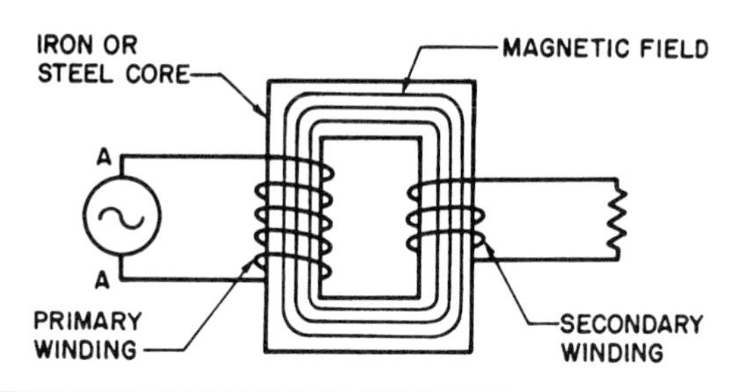

Figure 6-23 An illustration of a simple step-down transformer.

Figure 6-23 shows a basic step-down transformer. When voltage is applied to the insulated wire that is wrapped around the transformer core, a magnetic field is produced around the core. On the secondary side of the winding, the insulated wire picks up the induced magnetic field. Since there are fewer windings on the secondary side, a lower voltage is produced.

When working with gas furnaces, the transformer will have a 120-volt primary and a 24-volt secondary. When working with packaged air conditioning or electric furnaces, the transformer will have a 240-volt primary and a 24-volt secondary.

Transformers are rated by their VA (voltampere) capacity. The VA capacity of a transformer represents its ability to do work on the secondary side. If the number of loads on the secondary side of the circuit is low, such as only a gas valve and fan relay in a gas furnace, a transformer with a rating under 20 VA can be used. When working with heavier loads on air conditioning and commercial refrigeration systems, such as contactors and multiple relays, a 40 VA transformer is used. In some cases, such as larger air conditioning systems and heat pumps, you may find a transformer with a 65 VA rating.

Sizing Transformers

When replacing a transformer, use a replacement with an equal or higher VA capacity. However, avoid using a transformer that is rated excessively far above the original because it may create an unnecessary hum.

CONTACTORS

HVAC-R systems use contactors to carry high voltage to compressors, and in some cases, to motors used in other applications. As the name implies, the contactor contains contacts that close to complete the circuit to a load.

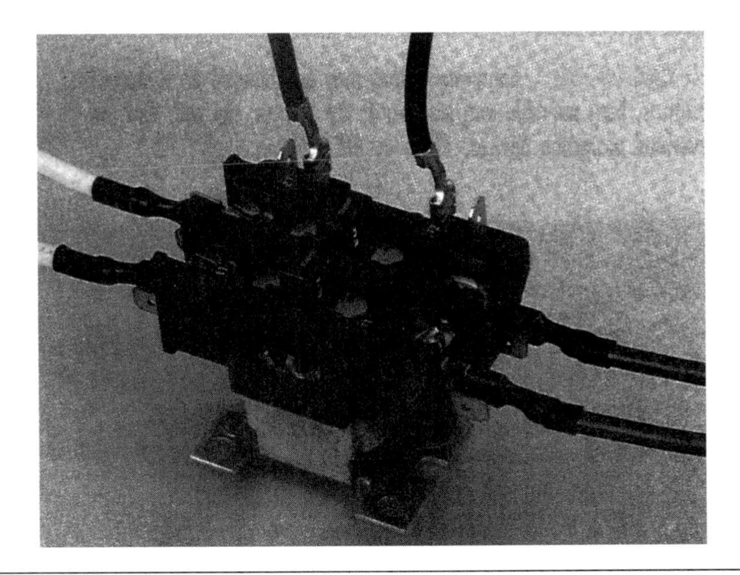

Figure 6-24 A contactor commonly used in HVAC-R electrical systems. (By Bill Johnson)

The method of operation of a contactor is based on electromagnetism. When the coil section of the contactor is energized, an electromagnetic field causes an assembly within the contactor to pull down, allowing the high voltage to bridge across the contact points. One type of contactor is shown in Figure 6-24.

The most common method of operating a contactor in an air conditioning system is a coil that operates on 24 volts. This allows a low voltage control circuit that ultimately applies high voltage to a heavy load. In some cases, contactors use coils that operate on higher voltages.

Contactors may be single pole (that is, breaking only one leg of power to a single-phase piece of equipment) or they may be two pole or three pole, depending on the application and the type of equipment.

Sizing Contactors

A good rule of thumb to follow when replacing contactors is to use replacement parts that are as close as possible to the original. For example, replacing a single pole contactor with a two pole contactor can create a problem. The manufacturer used a single pole contactor for a reason (usually to provide crankcase heat to the compressor during the off cycle) and your responsibility as a technician is to keep equipment as close as possible to its original design.

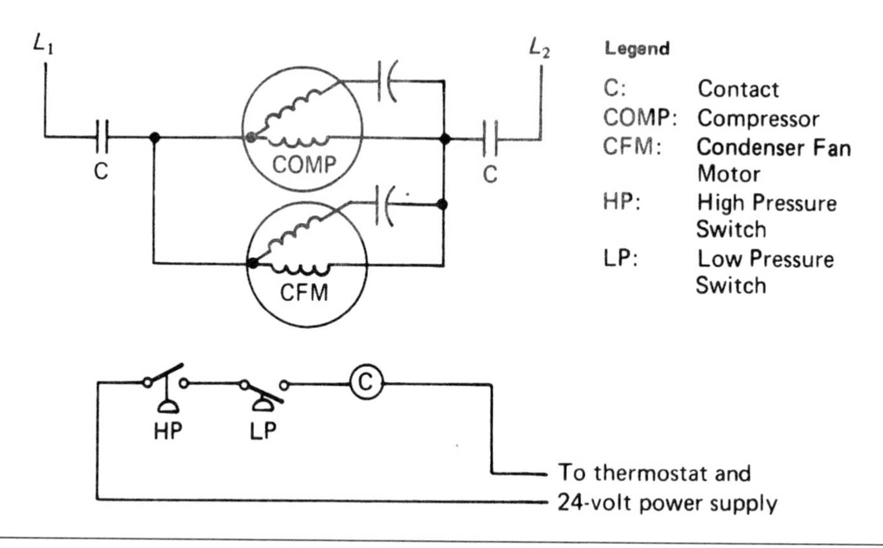

Figure 6-25 A schematic diagram of an A/C unit with a contactor controlling the compressor and outdoor fan motor.

Contactors are also rated by the amperage they are capable of carrying across their contact points (up to 60 amps). When replacing a contactor, always install a new one that is equal to or greater than the capacity of the original. For example, replacing a 40-amp contactor with a 30-amp contactor will only result in an early failure of the new component.

For an understanding of what a contactor does electrically, refer to Figure 6-25.

In many cases, when a contactor is shown electrically on a schematic diagram, half of the contactor is shown on the high voltage side of the circuit, and the other half is shown on the control voltage side of the circuit. (In residential and light commercial air conditioning equipment, the control voltage circuit is usually 24 volts.)

In the diagram shown, L1 and L2 are wired directly to the two sets of open contact points identified as C. On the control side of the diagram, the letter C also appears in a circle. This indicates the coil of the contactor.

The correlation between the C in the circle and the C's identified as contact points is as follows:

1. When power is applied to the low voltage circuit and the high pressure and low pressure switches are in the closed position, 24 volts will be applied to the coil.

2. When 24 volts is applied to the coil, the contact points will close and energize the compressor and condenser fan motor.

Why Contactors Fail

Contactors may be replaced for a variety of reasons:

- A break may occur in the coil wiring, resulting in an open coil that cannot provide a magnetic field to operate the armature that is connected to the contact points assembly.
- A coil may become shorted.
- The contact points may become badly burnt due to the repeated arcing that occurs naturally when the contactor makes a circuit to a load.
- A contactor may also be misdiagnosed and replaced by mistake if the voltage to the coil is not at the minimum required for operation. If a coil is supposed to operate on 24 volts and only 20 volts is applied, there will not be a strong enough magnetic field to pull down the armature assembly, and the contactor will "chatter," resulting in poor contact to the high voltage load.
- Insufficient voltage may be applied to a contactor coil due to a poor connection somewhere in the low voltage circuit or due to incorrect voltage applied to the primary of the transformer. If the correct voltage is not applied to the transformer primary, it cannot deliver the correct voltage from its secondary winding.

Study Questions

Now is a good time to take stock of what you have learned so far. Here are three study questions that refer to the diagram in Figure 6-25. Test your understanding of the material in this section.

1. True or false? The compressor and condenser fan motor are wired in a parallel circuit.
2. What type of motor is represented in the diagram?
3. Which winding is the start winding?

RELAYS

A relay operates in the same manner as a contactor in that a coil is energized through the control circuit to control another circuit to a load. While the contact points on a contactor are almost always normally open (NO) and are closed when the coil is energized, a relay will commonly have both NO contacts and NC (normally closed) contacts.

The terms *normally open* and *normally closed* refer to the position of the contact points when the relay coil assembly is not energized. With this feature used on more than one set of contacts, a relay can simultaneously control two separate circuits, making one when the coil is energized, and the other when the coil is deenergized.

A relay of this type is also used to control a multi-speed fan motor in a combination heating/cooling unit. In a common residential heating/cooling unit, the indoor fan motor runs on a lower speed for the heating mode, and a higher speed for the cooling mode. A relay is usually designed to carry a load up to 15 amps across its contact points. For a load with a higher rating, a contactor is used. Figure 6-26 shows some of the relays used in HVAC-R equipment.

For an electrical representation of a control relay, refer to Figure 6-27. In this diagram, a relay is shown controlling a multi-speed indoor fan motor. The diagram itself is built around the schematic symbol for a transformer. The control circuit is shown in the lower half of the diagram (24 volts) and the high voltage circuit is shown in the upper half of the diagram (110 volts).

Figure 6-26 A variety of relays used in HVAC-R equipment.

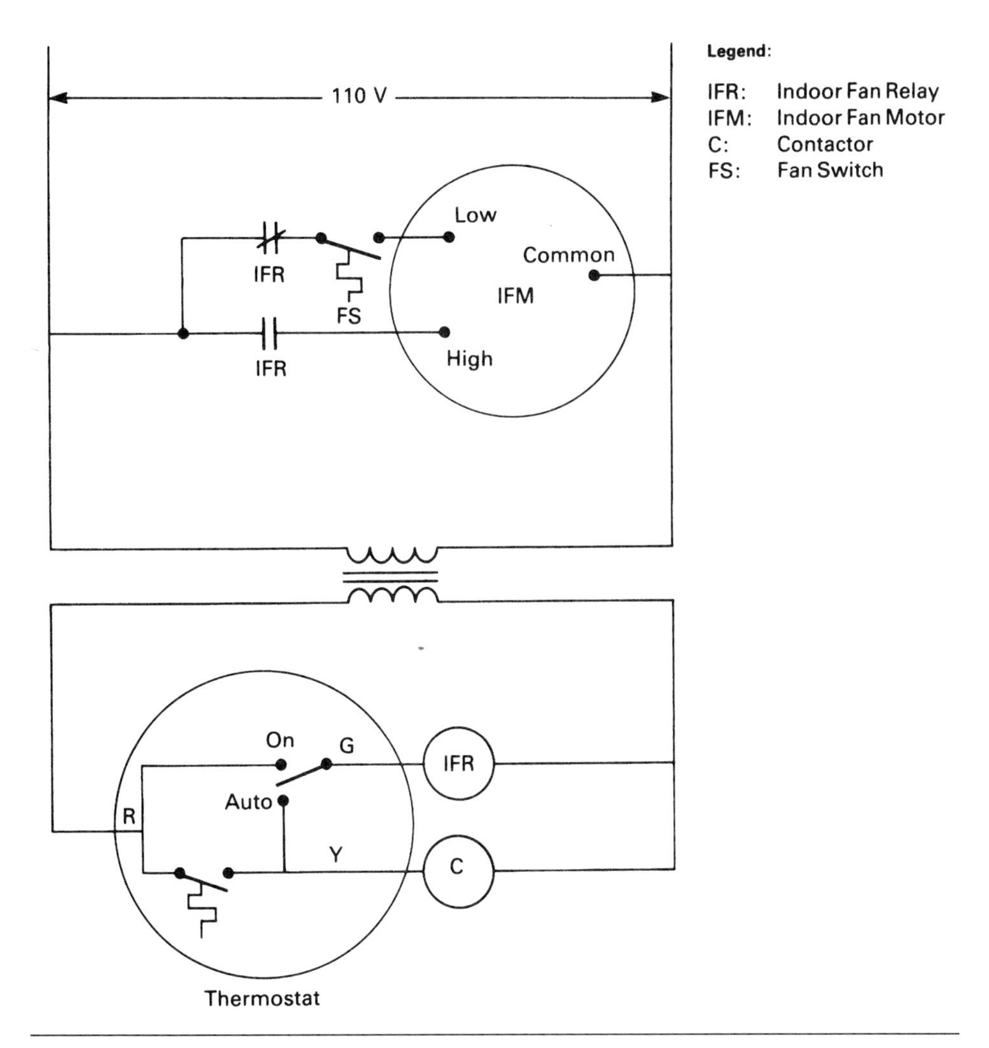

Figure 6-27 An electrical representation of a control relay.

Figure 6-27 shows another example of a coil with separate sets of contact points. In this circuit, the IFR (indoor fan relay) coil is indicated on the low voltage side, while the two sets of IFR contacts (one normally open and one normally closed) are identified on the high voltage side.

The specific type of motor in this diagram is not identified, but no capacitors or start devices are indicated. One side of the line (on the right of the schematic) is shown wired directly to the common terminal, while the other side of the line is wired to energize either the high speed or low speed of the motor, depending on the mode selected.

The high speed of this motor can be energized in one of two ways:

1. Turning the fan switch to the ON position will allow a complete circuit to the IFR coil, which will then close the IFR contacts wired in series with the high speed winding of the motor.
2. Placing the fan control switch in the AUTO position and allowing the thermostat to complete the circuit to the IFR coil. This also completes the circuit to a contactor coil, which would in turn allow the compressor and condenser fan (not shown) to be energized.

Keep in mind that when the IFR is energized, the normally closed contacts on the relay assembly will open, but in this discussion that fact is incidental since there is another switch in the circuit that is the primary control for the low speed on the fan motor. This switch, identified as FS, is shown as a *close-on-temperature-rise switch*. This type of switch is commonly used in a gas furnace. When the burners create enough heat in the heat exchanger section of the furnace, this switch senses the temperature rise and completes the circuit to the low speed winding of the fan motor. With this system, the fan in a gas furnace is energized after the burners have been on for a brief period of time.

Time Delay Relays

A time delay relay is used in several applications in HVAC-R equipment. This type of relay can be used to create a delay before the fan in a gas furnace starts. This allows the heat exchanger to warm up so that cold air is not blown into the conditioned space. It can also be used on a comfort cooling system to protect compressors from restarting too quickly in the event the customer turns the thermostat off and on repeatedly in the cooling mode. In addition, it can be used to accomplish what is known as a *part winding start* on large commercial system compressors. This allows only certain windings to be energized on the initial call for cooling, then allows other windings to be energized after a short delay. This is done to prevent excessive current surges when the motor is first energized.

Like the other control devices we have described, the time delay relay has a set of contact points that carry the voltage to the load, and also contains a control voltage section (usually some type of heating element) that is energized according to the system design. Two common categories of time delay relays are delay on make and delay on break.

Delay on Make – A delay on make relay is designed to create a delay when the control circuit is made to the control segment of the relay assembly.

This is the type of relay you may find in a gas furnace. For example, the control segment of the relay would be energized at the same time as the gas valve on the furnace, but would not allow the voltage to be carried to the fan motor until the designated delay time frame had elapsed.

Delay on Break – A delay on break relay assembly is designed to protect the compressor on an air conditioning or refrigeration unit in the event of a short-term power outage (or in the event the customer is playing with the thermostat). It will not allow the compressor to restart until the control voltage has been restored. Once the control circuit is complete again, the time delay assembly starts timing down to allow a restart.

TROUBLESHOOTING HINT

One way to identify certain time delay relays is to test between two of the four terminals on the relay with an ohmmeter. If you read some resistance between these two terminals, but no continuity between the two other terminals, you have located the load section of the relay. The two terminals without continuity are connected to the contact points within the relay and carry the high voltage to the fan motor or compressor after the load section has been energized for the prescribed period of time.

SEQUENCERS

Like a time delay relay, a sequencer prevents a load from being energized until a specified time lapse has occurred. A sequencer differs from a relay in that it may control two or more loads in a sequence, such as the heating elements in an electric furnace.

When a thermostat calls for heat from an electric furnace that contains two or more heating elements, a sequencer prevents the total load capacity from being energized simultaneously. This prevents an overload from occurring in the electrical supply system.

MAGNETIC STARTERS

A magnetic ("mag") starter is similar to a contactor in that it has a coil section that is energized by the control circuit, and a contact section that carries the operating voltage to the compressor. However, it differs from a standard contactor in that it contains a means of overload protection.

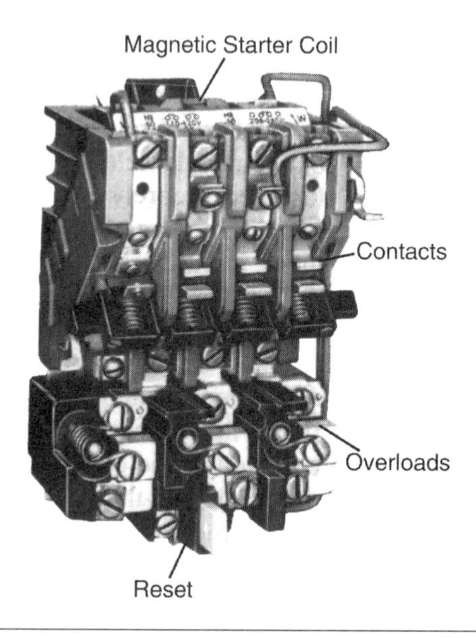

Figure 6-28 A "mag" starter used in large commercial systems.

Magnetic starters are used in large commercial refrigeration equipment that operates on three-phase power. Many starters are equipped with a manual reset (Figure 6-28). With this type of system, the unit restart can be monitored by a technician and the cause of the overload trip can be determined. A schematic illustration of a magnetic starter is shown in Figure 6-29.

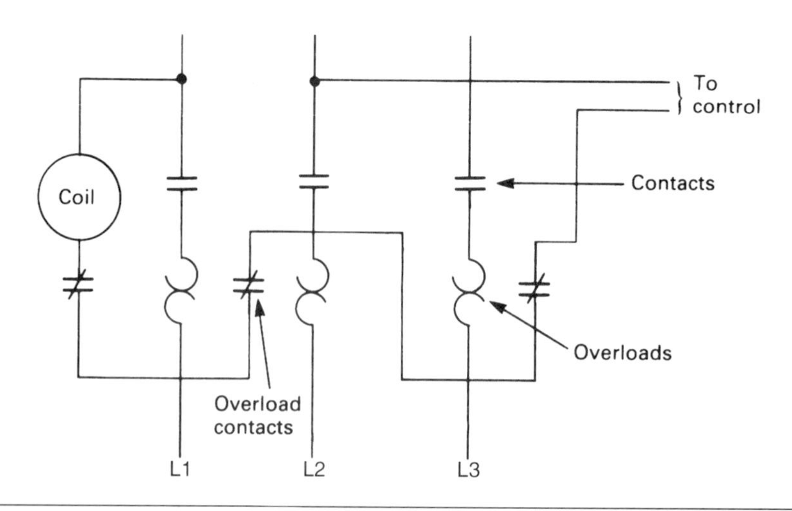

Figure 6-29 A schematic diagram of the "mag" starter.

In this schematic, the control voltage that operates the coil is drawn from L1 and L2 of the three-phase power. (L3 is not used.) L1 is routed directly to the operating coil, and L2 is routed to whatever control device is being used, then back through the normally closed overload contacts before being connected to the other side of the coil assembly.

PRESSURE SWITCHES

In order to protect and control the operation of some refrigeration systems commonly found in walk-in coolers, grocery store equipment, and restaurant equipment, a pressure switch is used. It controls the electrical circuit to the compressor. A pressure switch has terminals for connection of the electrical system and a flare fitting that allows connection to the appropriate section of the refrigeration system. Pressure switches have what are known as *cut-in points* and *cut-out points* that determine when the switch is activated.

Low Pressure Switch

Some systems use a low pressure switch such as the one shown in Figure 6-30. A low pressure switch connects to the low side of the refrigeration system and can be used in one of two ways: as a protective device

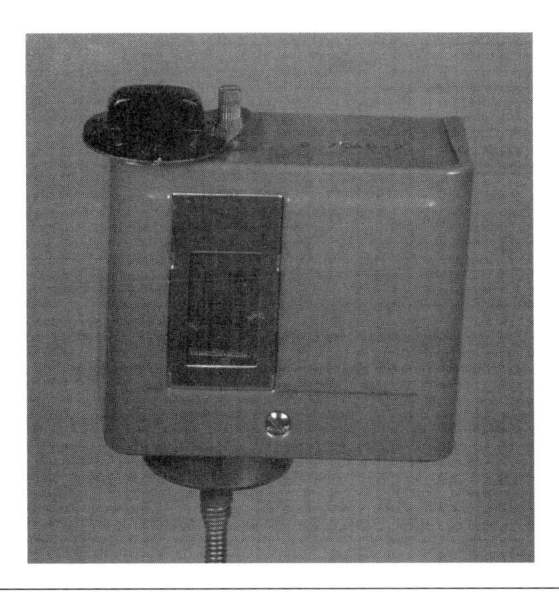

Figure 6-30 A low pressure switch can be used in a refrigeration system as a protective device and as a cycling device. (By Bill Johnson)

that will break the circuit to the compressor in the event of a loss of refrigerant, or as a device to control the on/off operation of the compressor.

When used as a protective device, the low pressure switch contacts will remain closed as long as there is sufficient refrigerant pressure on the low side of the system to overcome the spring pressure inside the switch. In the event of a refrigerant loss, the pressure decreases, and the spring causes the contacts inside the switch to break.

TROUBLESHOOTING HINT

In some cases, a small refrigerant loss will cause a pressure switch to shut a compressor down, but in the off cycle, the low side pressure will build up high enough to allow the switch to make, and the compressor will restart. However, it will only run for a short time, until the pressure again drops low enough to cause the switch to open and cut the power to the compressor.

The low pressure switch also works as a temperature control device because of the temperature/pressure relationship in the refrigerant. In other words, when the design temperature of the unit is reached, there is a corresponding pressure drop in the low side refrigerant pressure. The lower the temperature, the lower the pressure. During the off cycle, the temperature rises in the cooled space, and as the temperature rises, the pressure also rises. This rise in pressure causes the low pressure switch to make, and the circuit to the compressor is again established. With the refrigeration system in operation, the temperature drops again. When this occurs, a corresponding pressure drop takes place, opening the low pressure switch and cycling the compressor off again.

High Pressure Switch

A high pressure switch on a refrigeration system has only one job: to protect the compressor in the event of excessive pressure on the high side of the system. It is similar in appearance to the low pressure switch, but has a different range of cut-in and cut-out points. In some cases, you may find a low pressure/high pressure switch combination in use. One pressure sensing tube connects to the low side of the system, and the other pressure sensing tube connects to the high side of the system. Both switches are connected side-by-side on the bottom of the switch assembly.

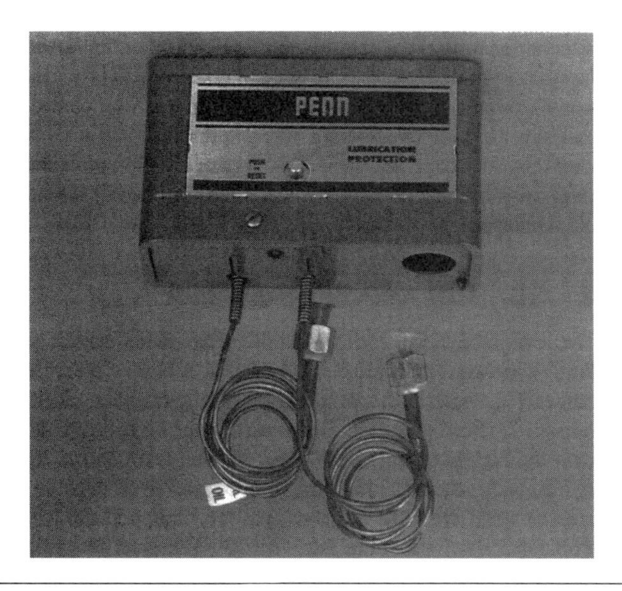

Figure 6-31 The function of an oil pressure safety switch is to measure net oil pressure.

Oil Pressure Safety Switch

An oil pressure safety switch also has two pressure sensing tubes, but they are not always connected on the bottom of the switch body like the high pressure/low pressure switch combination. Instead, the two sensing tubes may be opposite each other, one at the top of the switch and the other at the bottom of the switch. An oil pressure safety switch is shown in Figure 6-31.

The function of an oil pressure safety switch is to measure what is known as *net oil pressure*. Net oil pressure is the difference between the oil pump outlet pressure and the suction (low side) pressure of the refrigeration system. When one sensing tube is connected to the outlet side of the oil pump assembly in a refrigeration compressor, and the other sensing tube is connected to the suction side of the refrigeration system, the oil pressure safety switch is able to measure the net difference between the outlet pressure of the oil pump and the low pressure side of the system.

In the event of oil loss or inefficiency of the oil pump, the switch will break the circuit to the compressor until the shutdown can be detected and the problem diagnosed by a technician. Loss of oil or failure of the pump will result in a loss of lubrication on the crankshaft and bearing assemblies and cause damage to the compressor.

Figure 6-32 A typical defrost timer used in a domestic refrigerator.

TIMERS

Timers are used in both domestic refrigeration systems (frost-free refrigerators and freezers) and commercial refrigeration systems (low temperature walk-ins and reach-in refrigerators).

Domestic Refrigeration System Timers

An example of a timer found in frost-free refrigerators and freezers is shown in Figure 6-32. A cam inside the timer determines the position of the switch contacts inside the timer.

As shown in Figure 6-33, this type of timer energizes either the refrigeration system or the defrost system components. Tracing the neutral wire of the schematic, you can see that it provides one side of the circuit to the timer motor, defrost heater, compressor common terminal, and evaporator fan.

The hot leg of the 120-volt circuit is connected to the timer motor at terminal 1. In addition to providing power to the timer motor, it also applies power to either the defrost heater or the compressor and evaporator fan motor, depending on the position of the switches inside the timer.

This type of system is what is known as a *continuous run system* with the timer motor advancing toward the defrost mode whenever the refrigerator is plugged into the wall outlet. Continuous run systems are found on older refrigerators. On newer units, the timer is wired for what is known as *cumulative run time*. This means that the timer motor is wired in series with the control thermostat of the refrigerator, and the timer will advance toward defrost according to the amount of time the thermostat has called for cooling, rather than advancing toward defrost all the time.

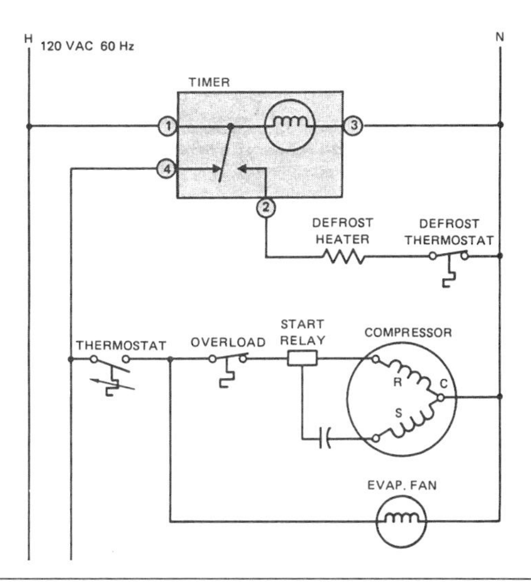

Figure 6-33 Schematic of a defrost timer used in a domestic refrigerator. In this application, the timer is wired for continuous run.

Commercial Refrigeration System Timers

Timers are also used in commercial refrigeration systems, such as frozen food display cases in grocery stores or walk-in freezers. This type of timer can be found controlling electric elements used for defrosting or energizing a solenoid that redirects the refrigerant flow in a hot gas defrost system. A typical commercial refrigeration timer is shown in Figure 6-34.

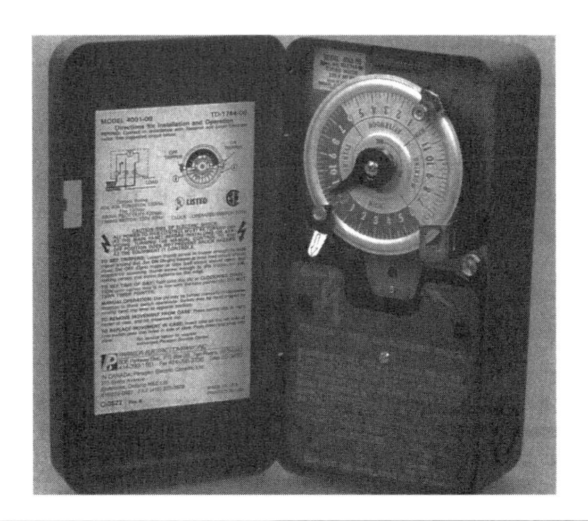

Figure 6-34 A timer used on commercial refrigeration systems. (By Bill Johnson)

UNIT SIX SUMMARY

Motors are used in HVAC-R equipment in a wide variety of applications: to operate fans in air handling systems, to provide energy for water pumps, and to operate control components such as defrost timers. The basis for the operation of all electric motors is magnetism. The two basic components in an electric motor are the stator and the rotor.

Split-phase motors are commonly used in HVAC-R equipment. This type of motor has two windings, a start winding and a run winding. In many cases, a split-phase motor start winding is dropped out of the circuit through the use of a centrifugal switch. A relay, such as a potential relay or current relay, may be used to provide a circuit to both windings at the instant of start, then break the circuit to the start winding after the motor has reached approximately 75 percent of its running speed.

A PSC (permanent split capacitor) motor uses a run capacitor wired in series with the start winding. Its purpose is to help the motor run more efficiently. In standard split-phase motors, the start winding is wound of much thinner and longer wire than the run winding, and therefore has a greater resistance. The windings in a PSC motor are not as different, and both windings are energized the entire time the motor is in operation. Many motors have internal overload protection while others, such as those used in fractional horsepower compressors, have external overload protection devices.

A step-down transformer is a component that allows a low voltage control circuit in an HVAC-R system. Most control circuits operate on 24 volts. In gas furnaces, a transformer is usually rated at 20 VA or under, while combination heating and cooling units in residences commonly use 40 VA transformers. Some larger heat pumps will use a 65 VA transformer.

Contactors are used to carry high voltage to a load and are controlled by a coil that is usually operated on a lower voltage. One pole, two pole, and three pole contactors may be used in different applications. A relay is similar in operation to a contactor, but is designed to control loads with a lower amperage rating. Fan relays are designed to carry up to 15 amps, while contactors are rated up to 60 amps. Time delay relays can be used in many applications, such as gas furnaces and as protective devices in comfort cooling systems. Sequencers are used to control the operation of heating elements in an electric furnace.

Magnetic starters are used in larger commercial refrigeration systems and differ from a contactor in that they have an overload system. Most magnetic starters will have a manual reset.

Pressure switches are used both as control and protective devices. A low pressure switch breaks the circuit to the compressor in the event of a loss of refrigerant. It can also be used to control the temperature in equipment such as walk-in coolers. A high pressure switch breaks the circuit to the compressor in the event of excessive high pressure. An oil pressure safety switch monitors the net oil pressure in a refrigeration system.

Timers may be used in refrigeration equipment to control the defrost cycle and the run cycle. Some timers may be wired for continuous run time, while others are wired for cumulative run time.

SECTION

3

TOOLS, EQUIPMENT, AND METERS

Tools and equipment used in the HVAC-R industry serve two purposes for the service technician. One is to aid in the troubleshooting process by evaluating the performance of equipment and diagnosing malfunctions. The other is to help make the repair once the problem has been identified.

This section describes the basic equipment necessary for troubleshooting refrigeration and electrical systems. With the wide range of tools and test instruments available, you will have to make many decisions on what types of test meters, gauges, and other equipment you want to invest in throughout your career. It is always wise to choose carefully and buy the best tools you can afford, but bear in mind that the tool is only as good as the technician who uses it. The most essential tool for any HVAC-R technician is a firm understanding of the components and operation of the equipment being serviced.

Refrigeration System Tools

Once you have isolated a problem in a piece of equipment as being within the sealed refrigeration system and decided what components you need to replace in order to get the unit back on line, you will need to know which tools to use. This unit discusses the tools required when piping, soldering, and evacuating sealed systems.

WORKING WITH TUBING

A basic fact you need to understand about tubing is the difference in the way some trades measure the material. For example, in plumbing and heating, tubing is sized by its inside diameter. In the HVAC-R industry, refrigeration tubing is measured by its outside diameter. Figure 7-1 shows the different methods of measuring type "L" tubing used in plumbing and heating and "ACR" tubing used in refrigeration systems.

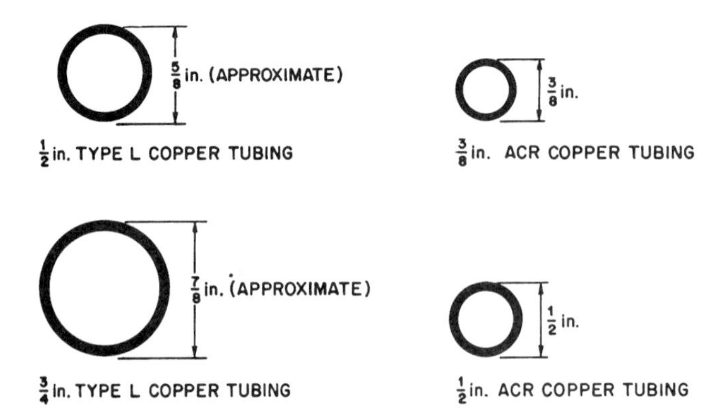

Figure 7-1 Different methods of measuring tubing.

What this means to you as a technician is that when you go into a supply house and ask for ⅝″ 90° elbows, the person behind the counter is going to ask if you want plumbing or refrigeration fittings. A ⅝″ 90° elbow would be ⅝″ to you as an HVAC-R technician, while the same fitting would be a ½″ 90° elbow to a plumber.

Refrigeration tubing itself comes in two forms: hard drawn (rigid and sold in 20-foot lengths) and soft, which is usually sold in 25-foot and 50-foot rolls.

When working with rigid tubing, you need to use various fittings, including couplers and elbows (ells). With soft copper, you can avoid the use of fittings in many cases.

Soft copper should be unrolled carefully so that you start out with straight pieces of tubing, then bend them in the shape you need. Figure 7-2 shows the proper method of unrolling soft copper tubing.

Figure 7-2 Soft copper tubing should be unrolled carefully so you start out with straight pieces of tubing. (By Bill Johnson)

Tubing Cutters

Common tubing cutters include those designed for tight spaces and smaller diameter tubing (up to ½″), and larger cutters used on tubing over ½″ in diameter. Figure 7-3 shows a small cutter and Figure 7-4 shows a larger cutting tool.

Larger tubing cutters may also be equipped with a reaming tool and an accessory known as a *restrictor wheel*. A restrictor wheel allows you to put a deep crimp in a large tube (soft copper) into which a smaller tube is positioned.

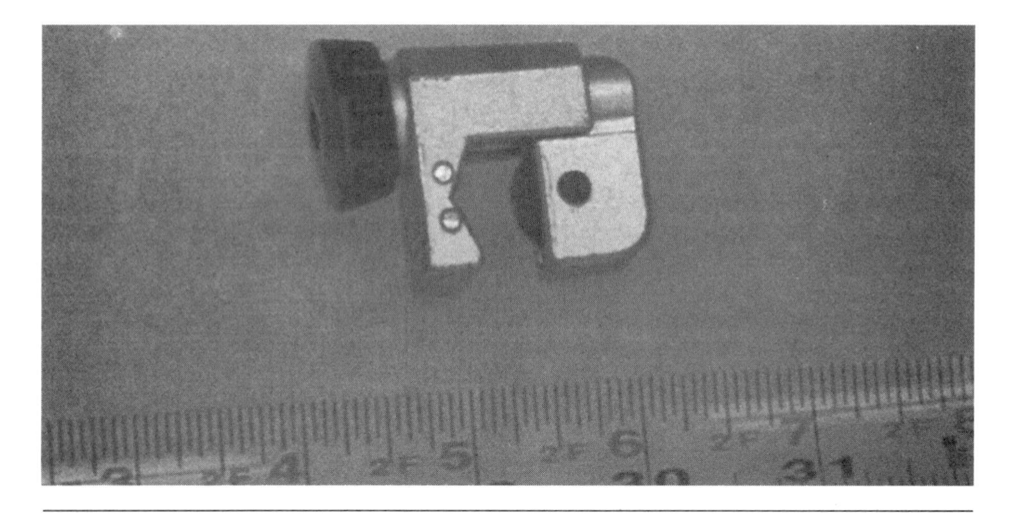

Figure 7-3 A small tubing cutter. (By Bill Johnson)

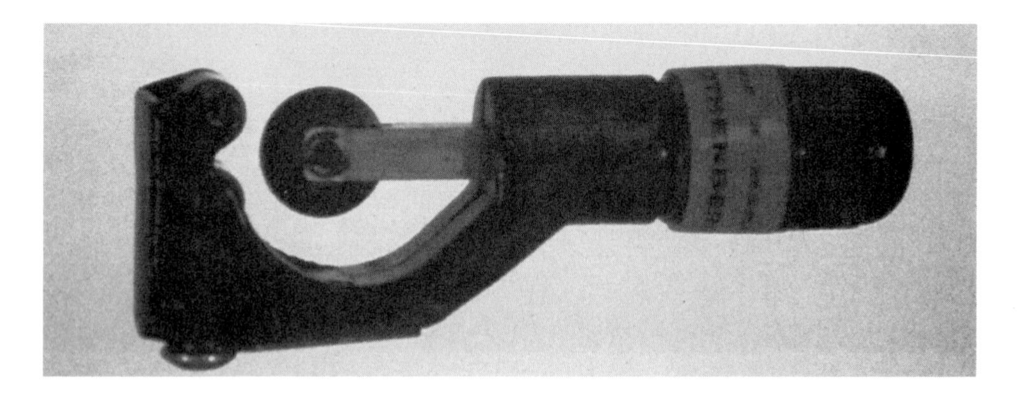

Figure 7-4 This type of tubing cutter may be equipped with a restrictor wheel. (By Bill Johnson)

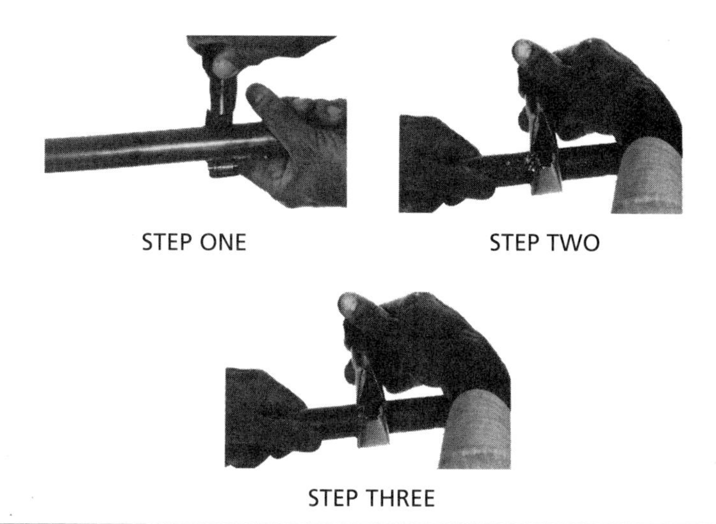

Figure 7-5 The proper way to use a tubing cutter. (By Bill Johnson)

Using a tubing cutter takes a little practice. Position the cutter as shown in Figure 7-5, then proceed as follows:

STEP ONE: Position the cutter on the tube by turning the adjusting wheel to a point where the tool is just snug on the tubing. Be careful not to turn it too tightly. Doing so may cause the tube to collapse or prevent you from being able to rotate the cutter.

STEP TWO: Rotate the tubing cutter around the tubing, tightening the adjusting wheel slightly with every couple of revolutions. Again, remember to tighten gradually.

STEP THREE: Keep turning and tightening until the cut is complete. After the tubing is cut, be sure to use a reamer to remove any burrs that may have formed during the cutting process.

TROUBLESHOOTING HINT

With some practice, you can learn to keep the burr that is created when cutting tubing to a bare minimum. The trick to this is to learn to cut the tubing almost all the way through, then snap the tubing at the cut rather than allowing the cutting wheel to penetrate all the way through the material. Any burr that is created must be removed with a reaming tool.

Preparing Tubing for Soldering

A good solder joint is the result of good preparation, which means cleaning the fitting and tubing properly, and removing all burrs. For proper tubing preparation, you need reamers, tubing benders, and tubing brushes.

Reamers – Reamers are available in several forms. Some of the most popular reamers are the inner and outer forms shown in Figure 7-6. To use a reamer, position it carefully and exert a small amount of pressure to accomplish removal of the burr either on the inside or the outside diameter of the tubing. Figure 7-7 shows a reaming tool being used.

Figure 7-6 Tubing reamers. (By Bill Johnson)

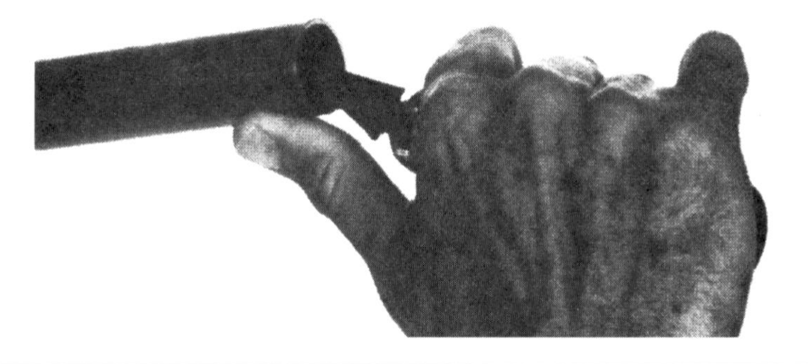

Figure 7-7 Using a reamer to remove a burr on a cut piece of tubing. (By Bill Johnson)

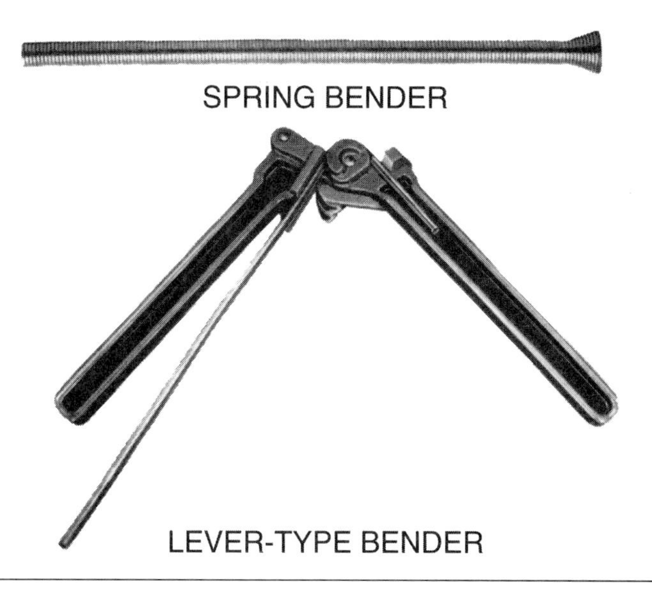

Figure 7-8 Two types of tubing benders. (By Bill Johnson)

Tubing Benders – Tubing benders are available in two basic types: the spring type and the lever type. Figure 7-8 shows both styles of benders.

Spring benders come in several sizes and must be selected to fit the diameter of the tubing used. If the spring does not fit tightly around the outer diameter of the tubing or fit snugly in the inner diameter, the tubing may collapse or the bend may be too sharp. Figure 7-9 shows the proper use of a spring bender.

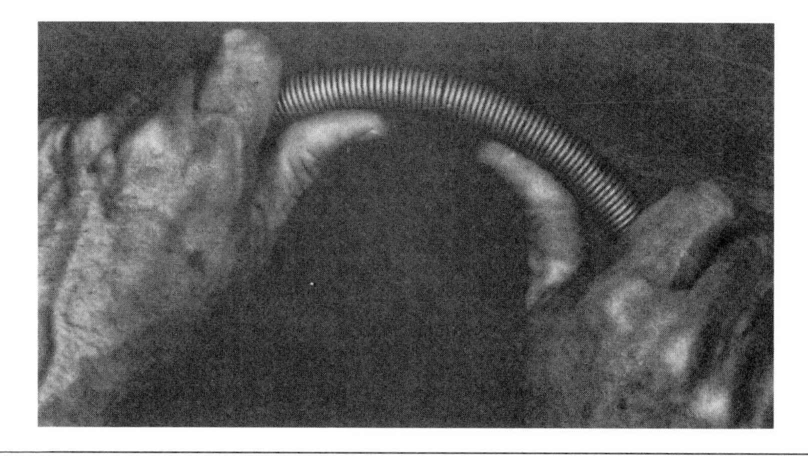

Figure 7-9 A bending spring is used to accomplish proper bends. (By Bill Johnson)

Figure 7-10 A lever-type bender being used to bend tubing. (By Bill Johnson)

Lever-type benders are usually available for a single size of tubing when used with tubing that is ⅝″ in diameter or larger. For smaller tubing, some benders are available for use with several sizes. Figure 7-10 shows a lever-type bender.

Tubing Brushes – Using tubing brushes when soldering joints helps to ensure a clean joint to which solder will adhere easily. Tubing brushes, such as those shown in Figure 7-11, are an effective way to remove dirt, grime, and the dull film that forms as a result of oxidation when tubing is exposed to the atmosphere.

Figure 7-11 Tubing brushes. (Courtesy Shaefer Brushes)

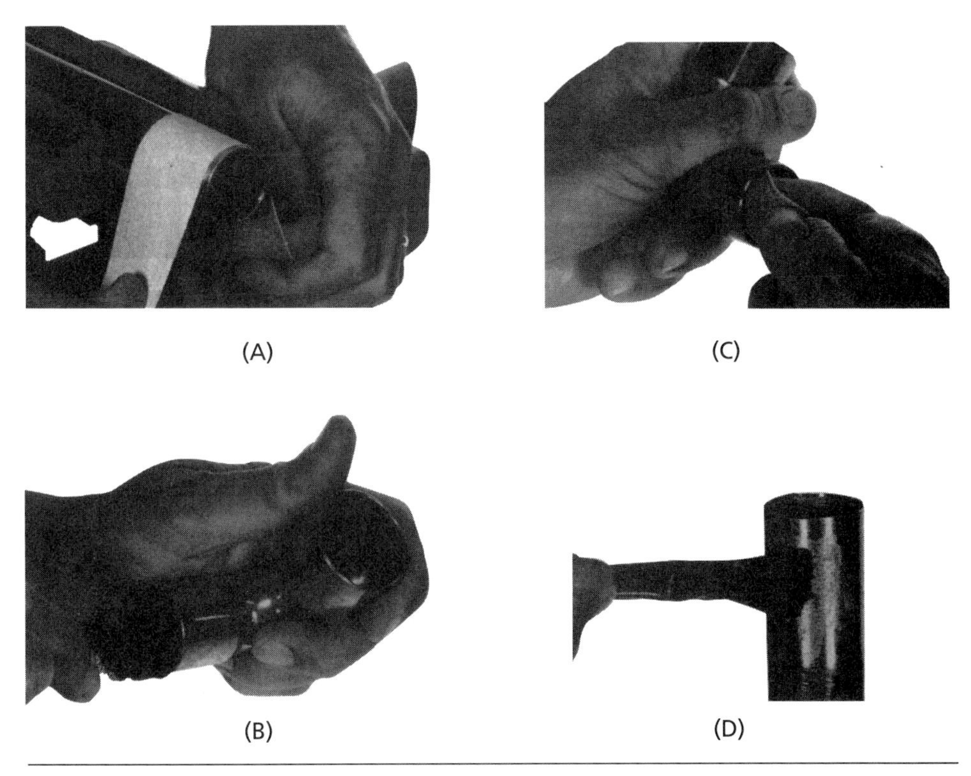

(A) (C)

(B) (D)

Figure 7-12 The proper procedure for cleaning and applying flux to tubing. (A) Clean with sand cloth. (B) Clean with a brush. (C) Use sand cloth on a fitting. (D) Apply flux. (By Bill Johnson)

A fine grit sand cloth can also be used to clean tubing in preparation for soldering. After you clean the tubing, avoid handling the prepared areas because oil from your hands will leave deposits on the clean tubing. Apply flux to the areas that will be joined. Figure 7-12 shows the cleaning and fluxing process.

Swaging Tools

Swaging tools allow you to make soft copper tubing connections without the need for a fitting and second soldered joint. Swaging tools are usually purchased in sets, with various sizes of tools for the different sizes of tubing used in refrigeration systems. A swaging tool is a tapered, punch-like tool that is used in conjunction with a block. The block holds the tubing while you drive the swaging tool down into the tubing to accomplish the swage.

To create a swaged joint, position the tubing in the block as shown in Figure 7-13, then insert the swaging tool (Figure 7-14). As shown in Figure 7-15, a ball-peen hammer is used to drive the swaging tool down into the tubing.

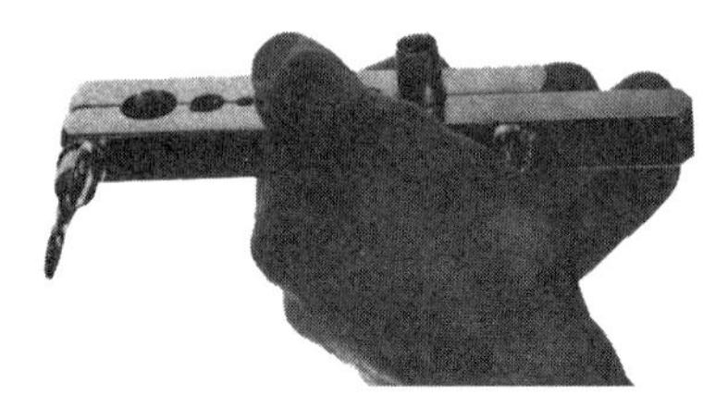

Figure 7-13 To accomplish a swaged joint, position the tubing properly in the block. (By Bill Johnson)

Figure 7-14 Inserting the swaging tool. (By Bill Johnson)

Figure 7-15 Using a ball-peen hammer to drive the swaging tool down into the tubing. (By Bill Johnson)

TROUBLESHOOTING HINT

It is not necessary to use a lot of force to accomplish a swaged joint. Use a reasonably light swing of the hammer to drive the swaging tool, and plan on making a few unattractive swages until you get the hang of driving the swaging tool straight down into the tubing.

A swaging tool set is also helpful when you need to determine the size of a piece of tubing. Tubing sizes are shown on the block.

Swaging tool sets usually include a flaring cone for making flared connections. Figure 7-16 shows the method of using the block and flaring cone to accomplish the flaring of the tubing.

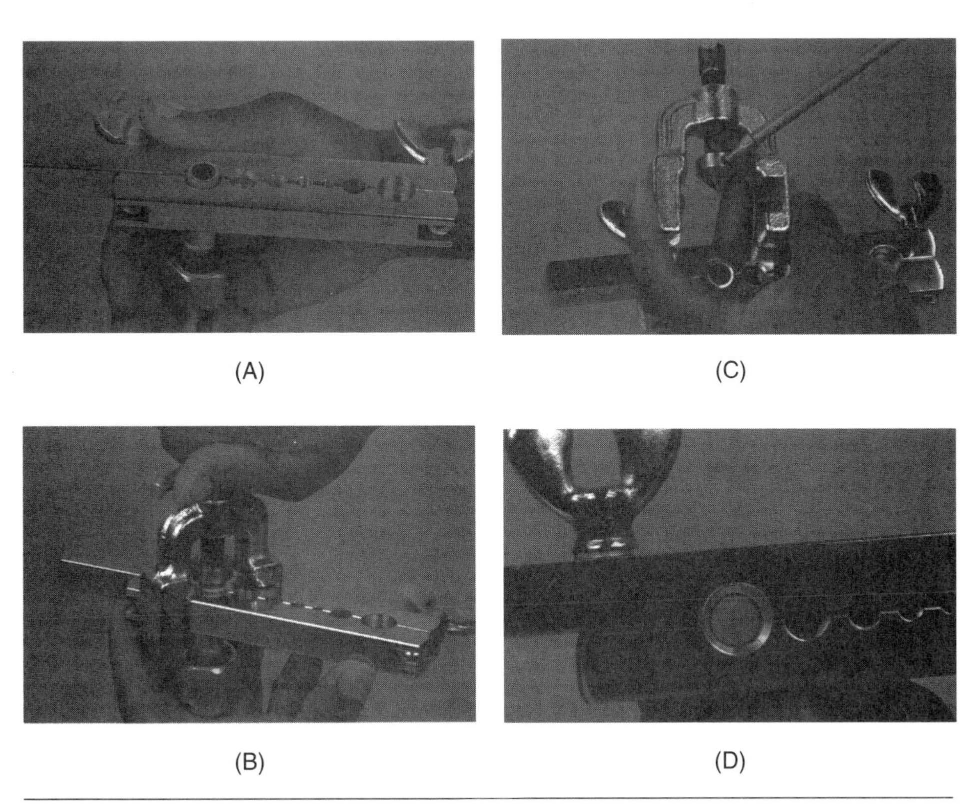

(A) (C) (B) (D)

Figure 7-16 The proper method of using a flaring tool. (A) Clamp the tube in the flaring block. Adjust it so that the tube is slightly above the block. (B) Place the yoke on the block with the tapered cone over the end of the tube. Lubricate with a drop or two of refrigerant oil, if desired. (C) Turn the screw down firmly. Continue to turn until the flare is completed. (D) Remove the tubing from the block and inspect for defects. (By Bill Johnson)

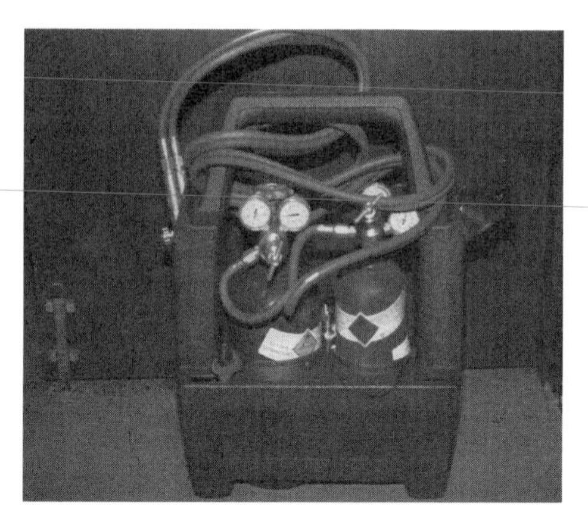

Figure 7-17 An air acetylene unit and an oxygen-acetylene unit. (By Bill Johnson)

Using Torches

Two types of torches used in the HVAC-R industry are the acetylene torch and the oxygen-acetylene combination torch. Both types are shown in Figure 7-17.

WARNING

Torches can be dangerous and when not used properly, can cause serious injury or death. Working with torches means working with tanks that are pressurized, and handling a dangerous open flame. Never use any torch assembly without being familiar with the safety practices that apply. The best approach to take when learning how to use a torch is to work with and learn from an experienced technician.

Once the tubing is cut, fitted, and fluxed, heat is applied with a torch to accomplish the joint. The most popular type of solder used in HVAC-R piping is known as *Sil-Phos*. Depending on the application, the silver content of this solder varies from a low of 1% to a high of 15%. Sil-Phos is classified as a high temperature solder with a melting point near 1,150°F.

Some applications use low temperature solders, such as 95/5 solder for some refrigerant system piping or 50/50 solder for water lines. Low temperature solder typically melts at 350°F.

When working with smaller domestic refrigeration systems, you may use silver solder, another high temperature alloy, which can have a silver content of up to 40%.

Whatever the solder, the technique for heating the tubing and getting the solder to flow properly into the joint is the same:

- Heat the tubing, not the solder.
- When the solder reaches its "plastic range," it will flow and fill in the joined area of the two fitted components. Figures 7-18 and 7-19 show the soldering procedure.

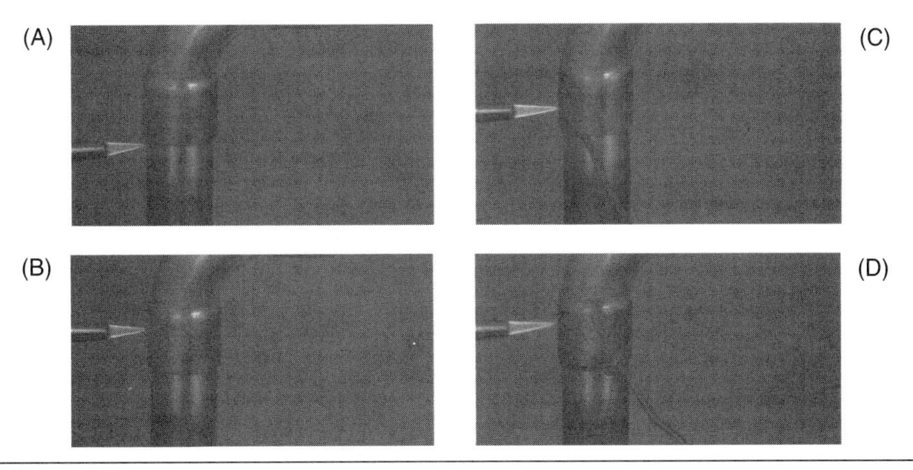

Figure 7-18 Proper procedures for heating a joint and applying solder. (A) Start by heating tubing. (B) Keep moving flame. Do not point flame into edge of fitting. (C) Touch solder to joint to check for proper heat. Do not melt solder with flame. (D) When joint is hot enough, solder will flow. (By Bill Johnson)

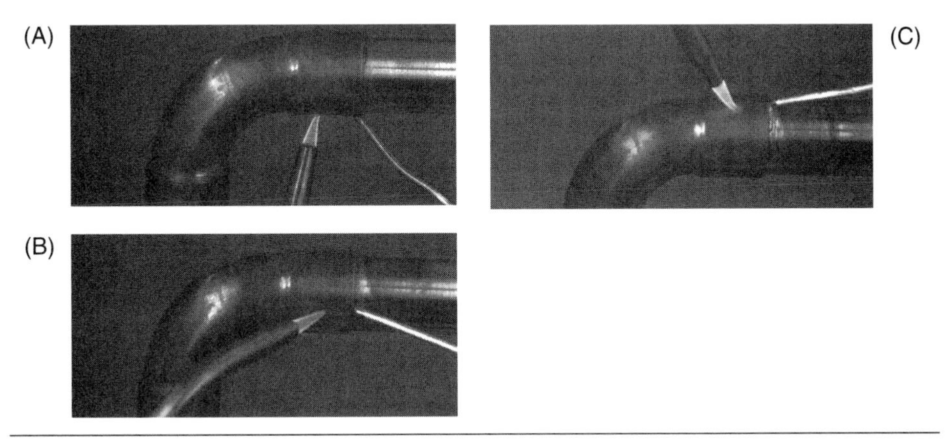

Figure 7-19 Making soldered or brazed joints in the horizontal position. (A) First apply the filler metal at the bottom. (B) Next, apply to the two sides. (C) Finally, apply to the top, making sure the operations overlap. (By Bill Johnson)

Figure 7-20 Vacuum pump. (Courtesy Robinair Division, SPX Corporation)

VACUUM PUMPS

After making a repair or part replacement on a sealed system, it will be necessary to evacuate the system before recharging with refrigerant. To accomplish this procedure, a quality vacuum pump that is capable of achieving a deep vacuum is required. To accomplish a deep vacuum, you need a rotary vacuum pump, preferably one that is two-stage. Figure 7-20 shows one type of vacuum pump available.

TROUBLESHOOTING HINT

Technicians who use home-made vacuum pumps that are basically nothing more than a refrigerator compressor are asking for trouble. Proper system cleanup, evacuation, and dehydration cannot be accomplished with such an item. A piston compressor, for example, is only capable of pulling a maximum vacuum of 28 inches. This is far from adequate for proper evacuation. A two-stage rotary vacuum pump is capable of pulling a vacuum close to the 29.94 inches that would be considered to be a perfect vacuum.

REFRIGERANT RECOVERY EQUIPMENT

With the advent of the Clean Air Act of 1990 and EPA regulations prohibiting the venting of refrigerants, recovery equipment has become an essential item on a service technician's equipment list.

Figure 7-21 This recovery/recycle unit is heavy and has wheels to make it easy to move around. (Courtesy Robinair Division, SPX Corporation)

While bulky recovery equipment is acceptable for shop work, field technicians need portable equipment in order to maintain compliance with refrigerant recovery laws. Figure 7-21 shows a recovery unit suitable for shop work, while Figure 7-22 shows a portable recovery system. In addition to the two-piece unit shown, other types of portable equipment are available.

Figure 7-22 This modular unit reduces a heavy unit to two manageable pieces. (Courtesy Robinair Division, SPX Corporation)

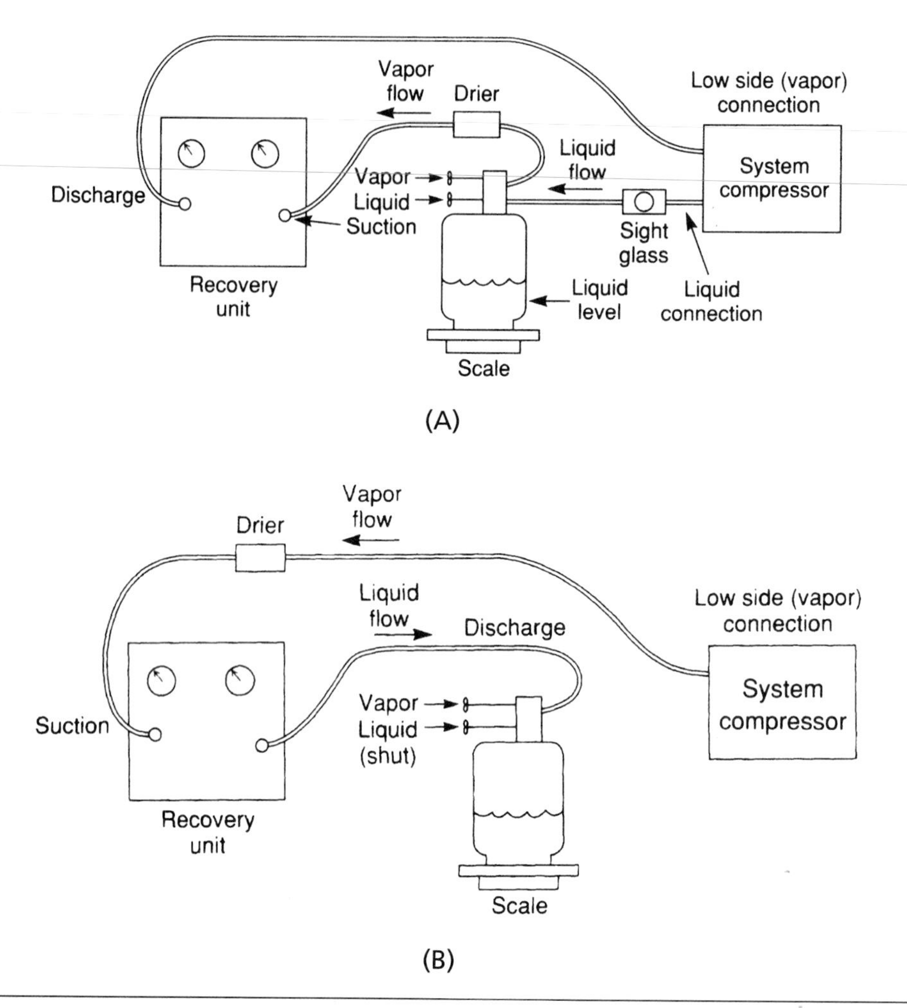

Figure 7-23 Two methods of recovering refrigerant from a system are the liquid method (A) and the vapor method (B).

Two methods of recovering refrigerant from a system are the *liquid method* and the *vapor method*. The vapor method is the slowest because refrigerant vapor is less dense than liquid, and takes longer to remove from the system. Figure 7-23 shows both methods.

The necessity for recovering refrigerant rather then venting it to the atmosphere has resulted in increased costs for customers, and as a professional, you have an obligation to keep these costs to a minimum. One way to speed up the recovery process and limit this cost is to keep as much heat on the system as possible while recovery is taking place (Figure 7-24).

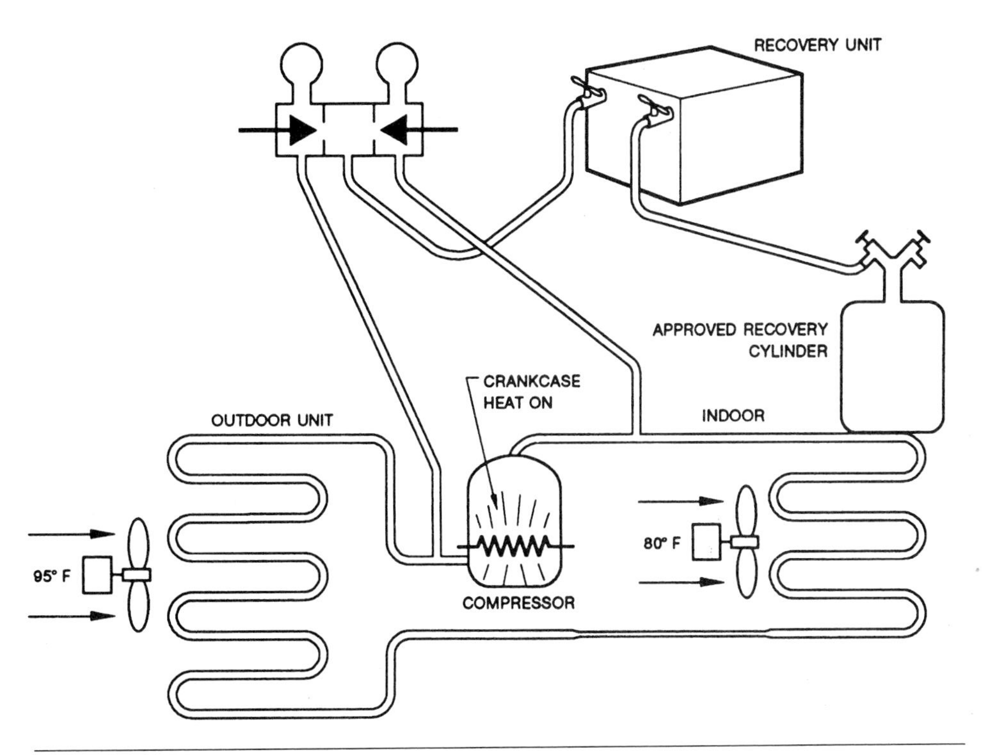

Figure 7-24 When recovering refrigerant from a system, keep as much of the heat on the system as possible. Keep the condenser fan and the evaporator fans on. Keep the crankcase heat on the compressor.

For example, energizing the crankcase heater, leaving the condenser and evaporator fans on, or using a heat gun to carefully warm the system coils are some of the ways you can accomplish the recovery process in as timely a manner as possible.

ACCESS VALVES

Access valves allow you to enter a refrigeration system to perform evaluation and service functions.

Smaller domestic systems (such as refrigerators) do not have factory-installed access valves, and they must be field-installed when necessary. Larger equipment, such as comfort cooling systems and those refrigeration systems found in grocery stores and restaurants, will have access valve assemblies installed at the factory.

Access valves on commercial equipment (shown in Figure 7-25) allow for isolation of the compressor as well as diagnostic and service procedures.

When working with access valves, use a service wrench such as the one shown in Figure 7-26.

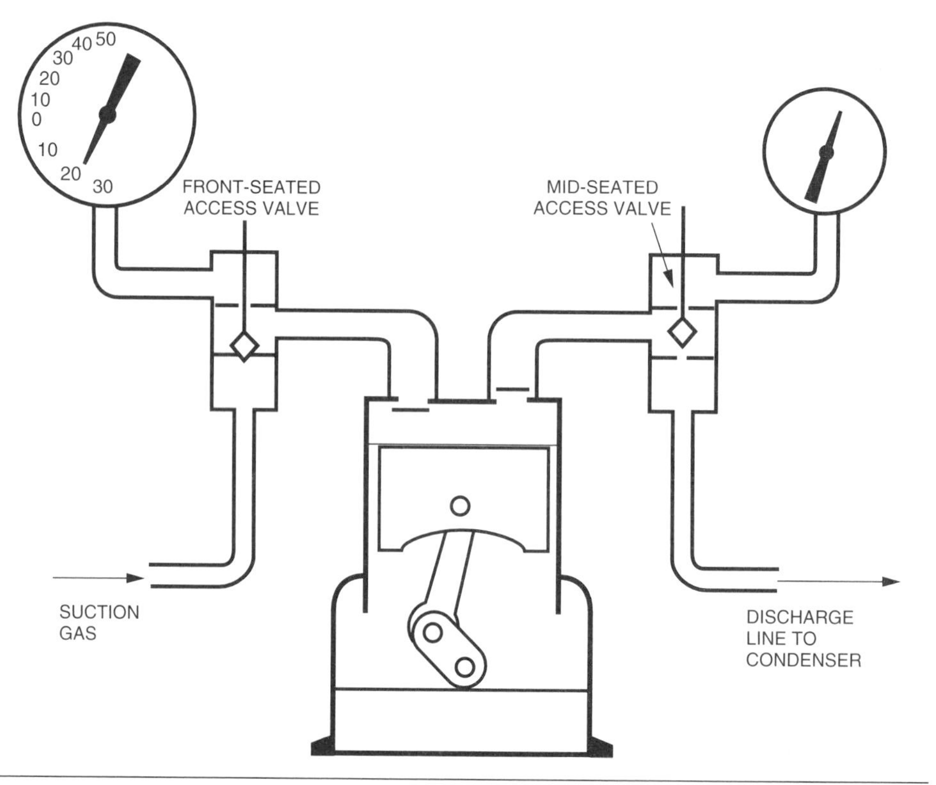

Figure 7-25 Commercial refrigeration system access valves.

Figure 7-26 A refrigeration service wrench. (Courtesy Robinair Division, SPX Corporation)

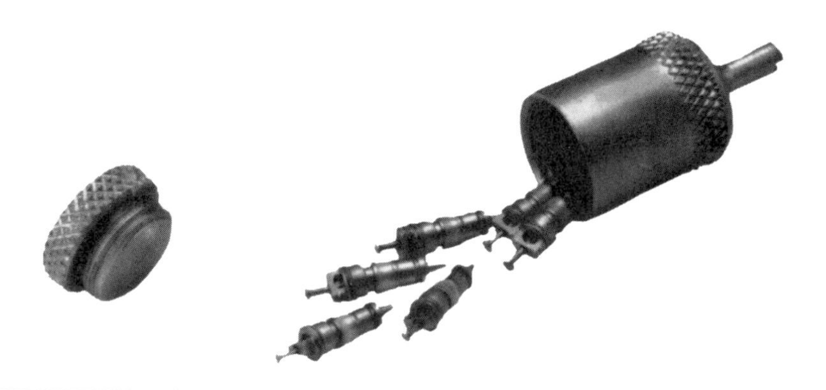

Figure 7-27 Spare Schrader valve cores along with a core removal tool. (Courtesy Robinair Division, SPX Corporation)

Schrader Valves

A common type of refrigeration system access valve is known as a *Schrader valve*. This type of valve has a spring-loaded core similar to that found in a bicycle tire. When a charging hose equipped with a fitting that depresses the core is used, it allows access to the system. It is a good idea to carry spare Schrader valve cores. They are available as a separate item or along with a standard core removal tool (Figure 7-27). A special core removal tool is also available that allows you to remove and reinstall the core while servicing the system (Figure 7-28).

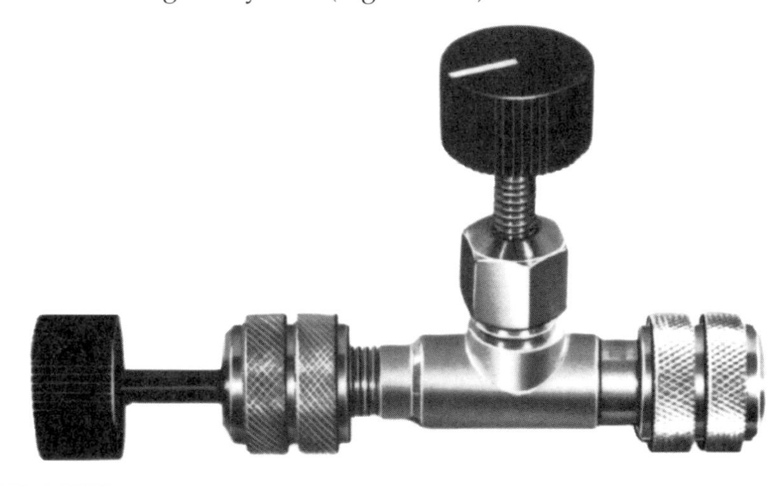

Figure 7-28 A core removal tool that allows you to maintain the integrity of a refrigeration system. (Courtesy Robinair Division, SPX Corporation)

When working with Schrader valves, you must perform service procedures quickly, and take all precautions necessary to prevent refrigerant discharge. One way to accomplish this is to use the special core removal tool shown in Figure 7-28. It allows removal of the core while still maintaining the integrity of the system. Figure 7-29 shows the difference in gas flow when the core is removed.

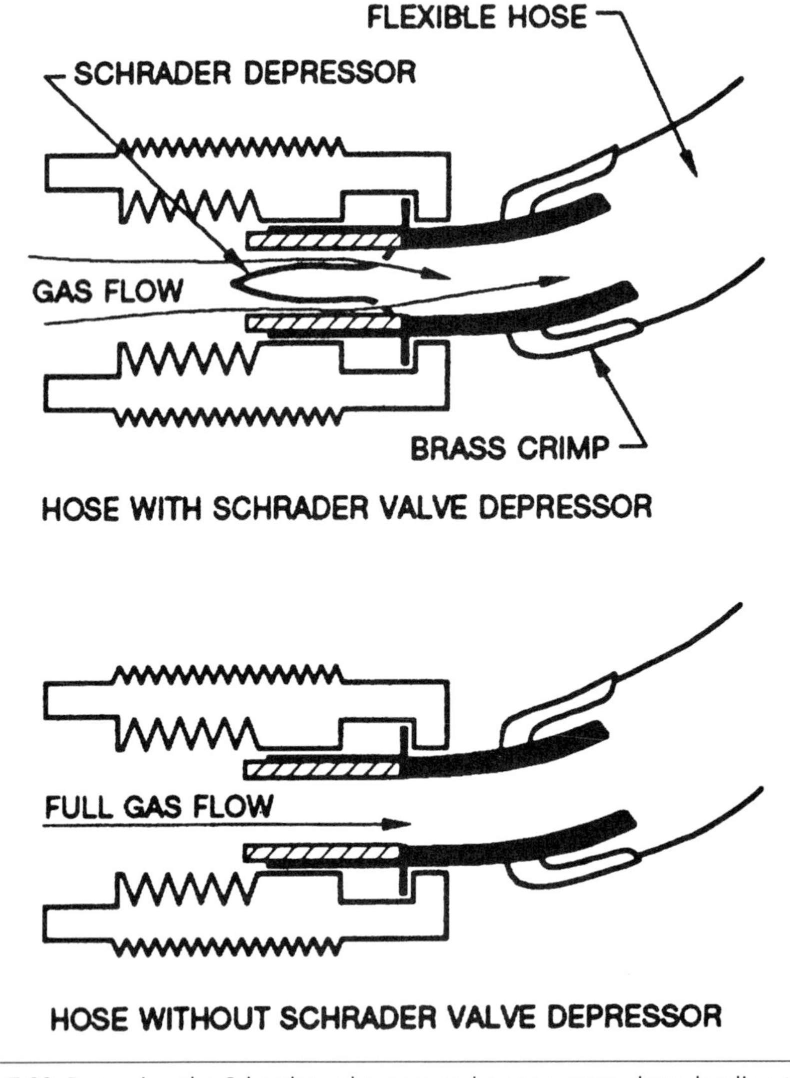

Figure 7-29 Removing the Schrader valve core reduces pressure drop, leading to less restriction.

CHARGING CYLINDERS AND SCALES

Many refrigeration systems are designated as *critical charge systems*. This means that the refrigerant charge must be measured precisely. Any overcharge or undercharge will affect the system's operating efficiency. To accomplish proper refrigerant charging, you may use a charging cylinder, such as the one shown in Figure 7-30, or an electronic scale, such as the one shown in Figure 7-31.

Figure 7-30 Charging cylinder. (By Bill Johnson)

Figure 7-31 Electronic charging scale. (Courtesy Robinair Division, SPX Corporation)

REFRIGERATION SYSTEM DIAGNOSTIC TOOLS

Compound Gauges

Gauges are one of the most important diagnostic tools available to the HVAC-R technician. A common type of gauge set is the three-hose type that technicians have used for many years. Four-hose sets have also been available for some time, but have only come into popular use as a result of the refrigerant recovery laws. Figure 7-32 shows the two types of gauge sets. Commonly referred to as *compound gauges* or a *manifold set*, one gauge is used to indicate low side pressure and the other is used to indicate high side pressure.

The gauge on the left is used to monitor the low pressure side and is often referred to as a compound gauge itself because it measures both pressure and vacuum (negative pressure). The gauge on the right is used to monitor high pressures. A common method of color coding the hoses is to use red for high pressure, blue for low pressure, and yellow for the center hose.

With a basic set of gauges, the hose on the right is connected to the high pressure side of the system and the hose on the left is connected to the low pressure side of the system. The center hose can then be connected to

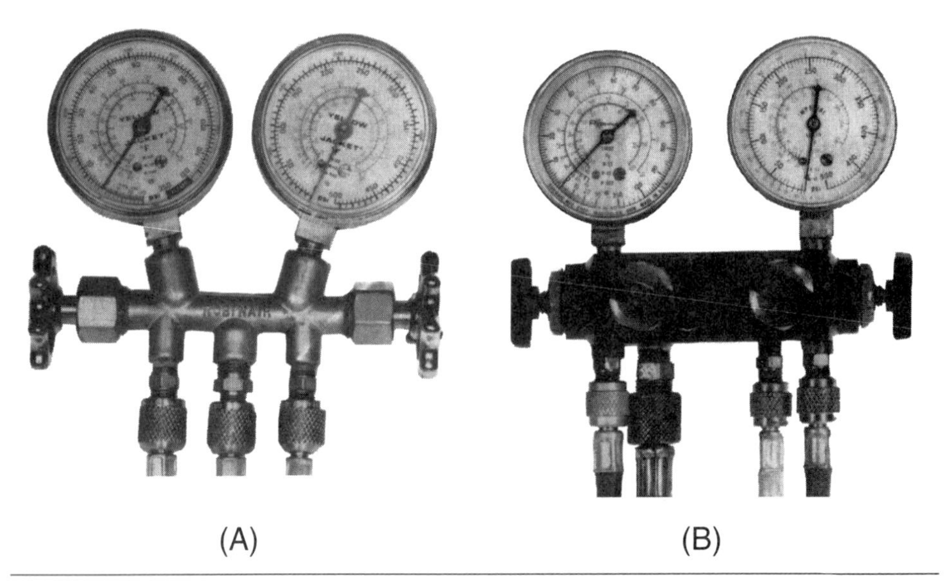

Figure 7-32 Gauge manifolds. (A) Two-valve gauge manifold. (B) Four-valve gauge manifold. (By Bill Johnson)

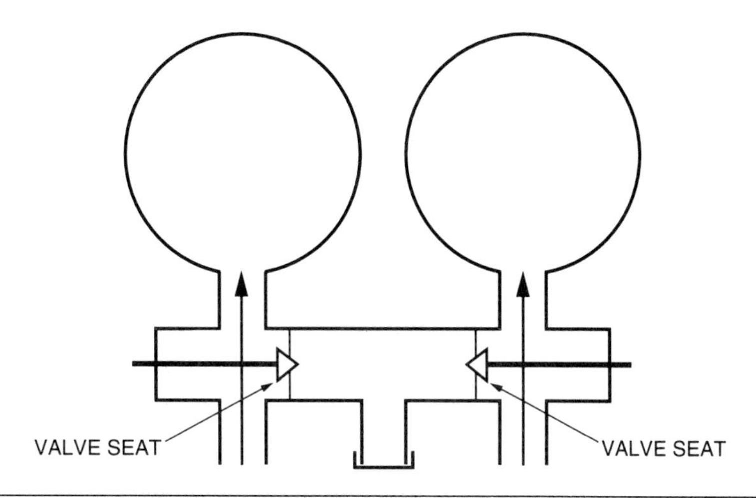

Figure 7-33 The valve seat on a gauge set allows you to read the gauges with the hand valves in the closed position.

a vacuum pump, refrigerant recovery system, or refrigerant container when evacuating and charging a refrigeration system. In the event that you are simply monitoring the operating pressures rather than evacuating or charging a system, the center hose would not be used. In this case, it is fastened to the threaded stub located on the manifold.

In a gauge set, the seat on each of the valves is located as shown in Figure 7-33. This means that you can leave both hand valves in the closed position and still read the pressure in the system.

TROUBLESHOOTING HINT

You should not have both gauge valves fully open when reading the pressure in a system. When you connect both hoses to a system and open both hand valves, this provides a path for the high pressure side to bleed over onto the low side of the system, causing incorrect pressure readings.

In addition to the standard service gauges, more sophisticated equipment is also available. One example is a set of oil-filled gauges such as those shown in Figure 7-34. Oil-filled gauges (also referred to as *glycerin-filled*) are used to dampen the needle vibration that can be experienced with standard gauges. Oil-filled gauges are also better protected from sudden pressure surges. Because of the developments in the HVAC-R

Figure 7-34 Oil-filled gauges are protected from sudden pressure surges. (Courtesy TIF Instruments, Inc.)

industry regarding the use of refrigerant recovery equipment, gauges have become even more sophisticated, with more hose connections and valves used to accomplish refrigerant recovery and system charging.

Digital Meters

Another type of diagnostic tool available is a digital meter that is used with a special adapter to provide a direct readout of the operating pressure in a sealed system. Figure 7-35 shows one type of digital meter available.

Leak Detectors

Both halide and electronic leak detectors are commonly used to find leaks in the refrigeration system.

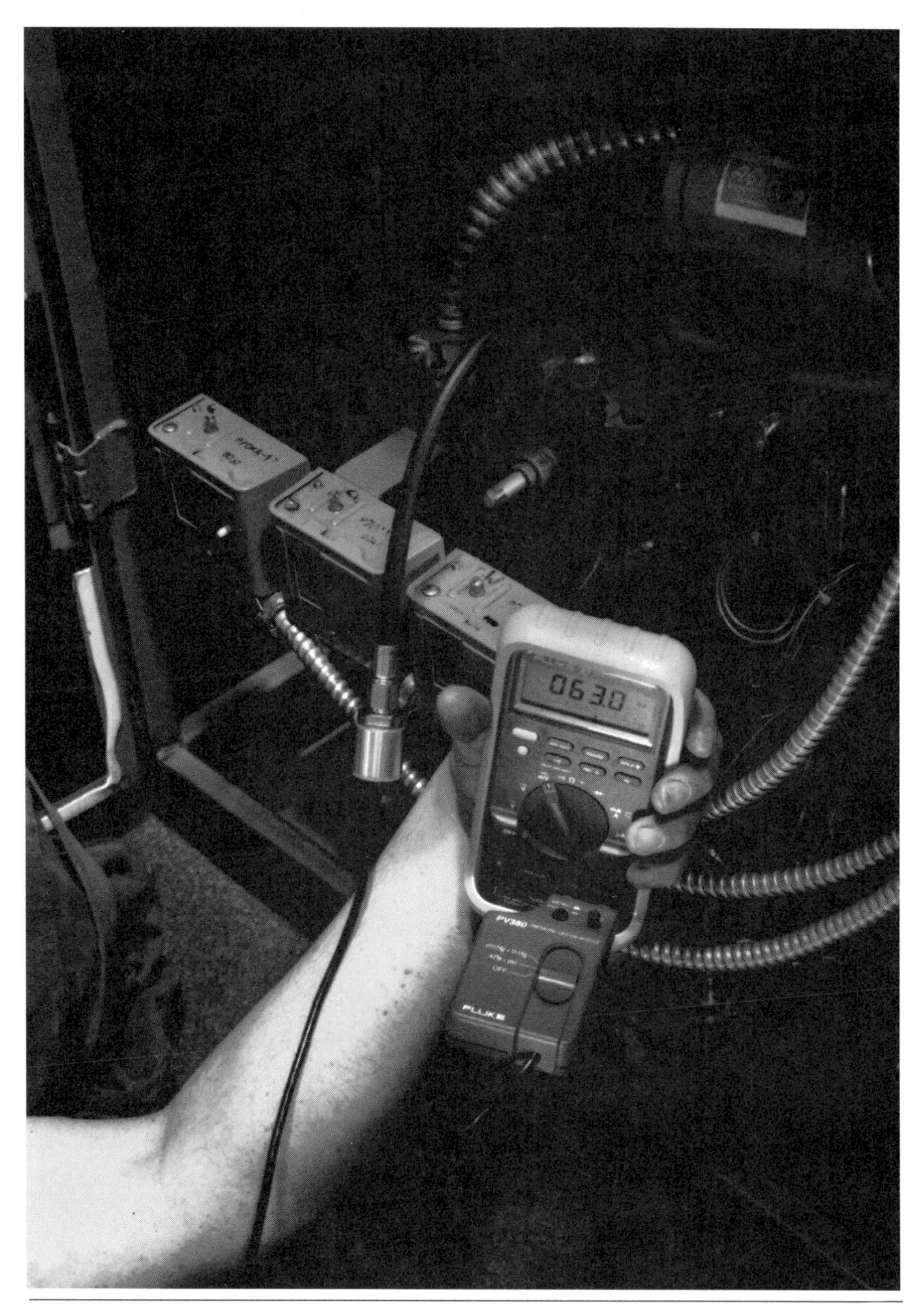

Figure 7-35 A meter that offers a digital readout of refrigeration system pressure. (Reproduced with permission of The Fluke Corporation)

Halide Leak Detectors – A halide leak detector, such as the one shown in Figures 7-36 and 7-37, is constructed of a copper disc that is heated by an open flame. When the refrigerant contacts the disc, it changes the color of the flame and a leak is indicated. For example, a small leak would be indicated by a green flame, while a purple flame would indicate a leak of larger magnitude.

Figure 7-36 Halide leak detector used with acetylene. (By Bill Johnson)

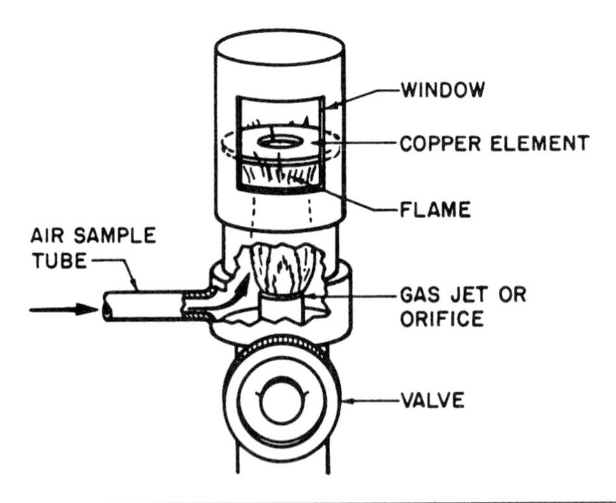

Figure 7-37 A halide leak detector is constructed of a copper disc that is heated by an open flame.

Electronic Leak Detectors – An electronic leak detector, often referred to as a *sniffer*, is a more sensitive leak detection device. Most can be adjusted to operate in contaminated areas so that they can find the actual location of the leak, rather than react to a room full of refrigerant and indicate a leak everywhere. Figure 7-38 shows two types of electronic leak detectors.

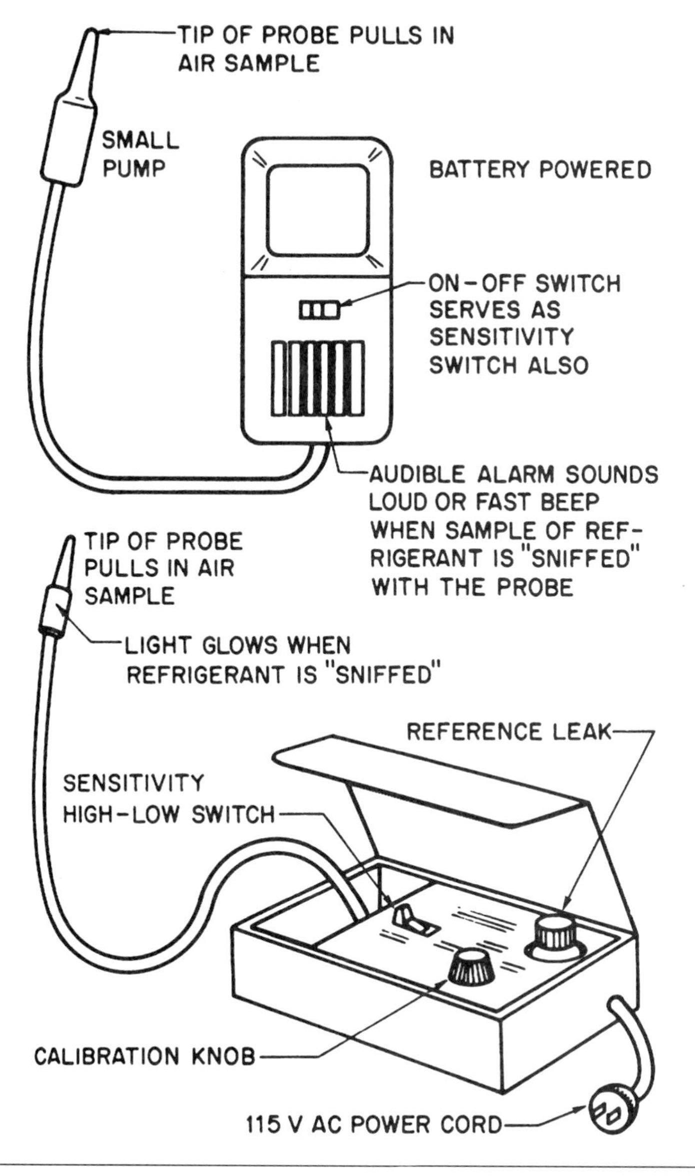

Figure 7-38 Electronic leak detectors.

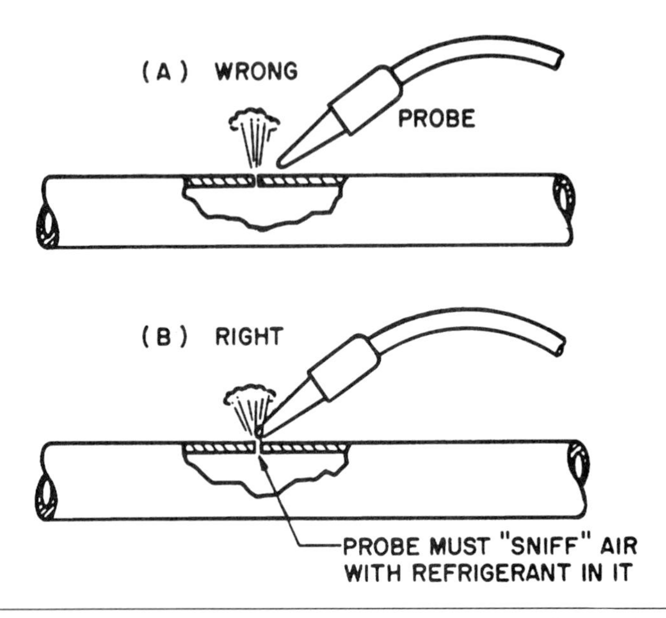

Figure 7-39 (A) Electronic leak detector is not sensing the small pinhole leak because it is spraying past the detector's sensor. (B) Sensor will detect refrigerant leak.

One sounds an audible alarm when a leak is found, while the other indicates leaks with a flashing light.

Figure 7-39 shows the method of operation of a typical electronic leak detector.

UNIT SEVEN SUMMARY

Refrigeration system tubing is measured by its outside diameter, as opposed to plumbing system tubing, which is measured by its inside diameter. Two types of copper tubing used in HVAC-R systems are hard drawn (rigid) tubing and soft copper tubing.

When cutting tubing, be careful not to crimp it by excessive tightening of the cutting wheel. Both large and small cutting tools are available.

Two types of tubing benders are the spring bender and the lever-type bender. To avoid tubing damage when bending, the correct size of bender must be used. Some lever-type benders may be designed to handle more than one size of tubing.

Swaging tools allow you to make soft copper tubing connections without the need for a fitting and a second solder joint. When equipped with a flaring cone, a swaging tool set tubing block can also be used to flare tubing.

Two common torches are the acetylene torch and the oxygen-acetylene torch. When using a torch, always exercise extreme caution. The open flame can cause damage or injury if safety precautions are not followed.

When soldering joints, always heat the tubing, not the solder. Proper preparation for soldering includes cleaning the tubing, reaming out any burrs, and using flux. High temperature solder melts at approximately 1,150°F and low temperature solder melts at approximately 350°F.

A rotary vacuum pump is necessary to pull a proper vacuum on a sealed system. Home-made vacuum pumps that use piston compressors cannot do a proper job of evacuating and dehydrating a sealed system.

With the advent of the refrigerant recovery laws, recovery equipment has become a necessity for all HVAC-R technicians. Heavy equipment is suitable for shop use, while portable equipment is needed for field work. To speed up the recovery process, keep as much heat on the system as possible without causing damage.

Access valves allow entry into the sealed system to accomplish evaluation and servicing. Smaller domestic refrigeration systems do not come equipped with access valves. Refrigeration equipment such as that found in grocery stores will have factory-installed access valves that also allow isolation of the compressor. Schrader valves are the most common type of service valve. They have spring-loaded cores that are similar to those found in bicycle tires.

When charging a refrigeration system, a charging cylinder or electronic scale can be used to make sure the correct charge is installed. When dealing with a critical charge system, an undercharge or overcharge will affect the operating efficiency of the unit.

A set of manifold gauges allows you to read the pressure in a sealed system. Digital meter adapters are also available to monitor system pressures.

Two common leak detectors are the halide and electronic types. A halide leak detector uses an open flame and copper disc to find leaks, while an electronic device uses a probe and either an audible or flashing alarm system.

Electric Meters

Troubleshooting the electrical circuits and components in HVAC-R equipment will make up approximately 75 percent of your work day. In order to properly evaluate and repair a malfunctioning electrical system, you have to be familiar not only with the electrical circuits and components, but also with the meters and other test equipment used in circuit monitoring and troubleshooting.

This unit provides an illustrative review of multimeters (VOM's), ammeters, temperature sensing devices, capacitor analyzers, and circuit testing devices. It also explains how to use a VOM.

MULTIMETERS

A multimeter is the test instrument most often used for tracking down problems in a malfunctioning system. Two types of multimeters are the analog multimeter, sometimes referred to as a *volt-ohm-milliammeter* or *VOM*, and the digital multimeter, sometimes referred to as a *DMM*. (See Figure 8-1.)

Digital meters have several advantages over the analog type. Among them are the fact that analog meters have to be adjusted to zero for resistance measurements. In addition, reading errors can result from reading an analog meter at an angle, a condition known as *parallax*.

Multimeters are frequently used to measure the voltage in a circuit, the resistance of a component, such as a coil or heating element, or the continuity across switch contacts and in wiring harnesses. While separate voltmeters, ohmmeters, and ammeters are available, most technicians prefer to use a combination meter, especially for measuring voltage and resistance.

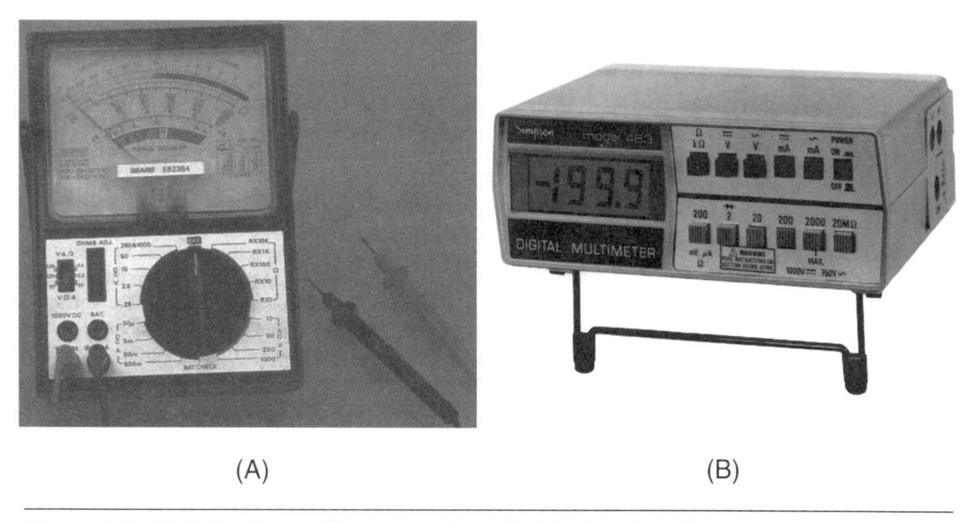

Figure 8-1 (A) Volt-ohm-milliammeter (VOM). (B) Digital multimeter. (Courtesy Simpson Electric Co., Elgin, IL)

AMMETERS

An ammeter is the test device used to check an electrical circuit to see how much current is flowing through the wiring. They are also used to test components for the correct amperage draw. Figure 8-2 shows various types of ammeters and how they are used.

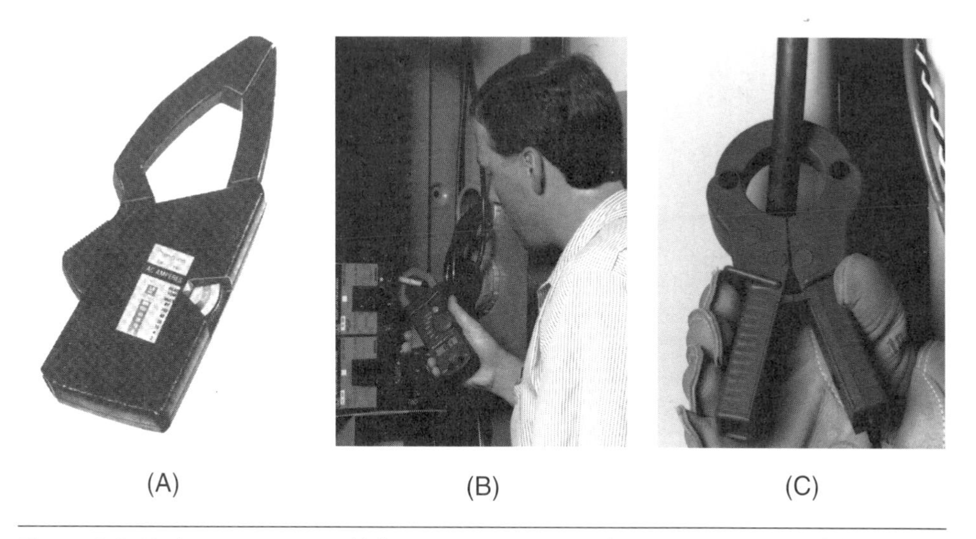

Figure 8-2 Various ammeters. [(A) Courtesy Simpson Electric Co., Elgin IL; (B) and (C) courtesy The Fluke Corporation]

Ammeters sense the magnetic field that surrounds all electrical conductors when current is flowing in a circuit. When using an ammeter, the jaws must be clamped around only one wire of the circuit to avoid canceling the reading.

WARNING

Always remember that you are working with live circuits, and exercise extreme caution to avoid touching "hot" wires. Wear heavy-duty rubber gloves and rubber-soled shoes when working near high voltage circuits.

KILOWATT HOUR METER/WATTMETER

A kilowatt hour meter/wattmeter is basically an energy management tool designed to test a circuit to determine how the energy is being used (Figure 8-3). You can also obtain accumulated power consumption in kilowatt hours over a period of time. This is helpful when trying to determine whether or not a piece of equipment is at fault for excessive electric bills.

A meter of this type can be used on both single-phase and three-phase systems. A three-phase connection is shown in Figure 8-3.

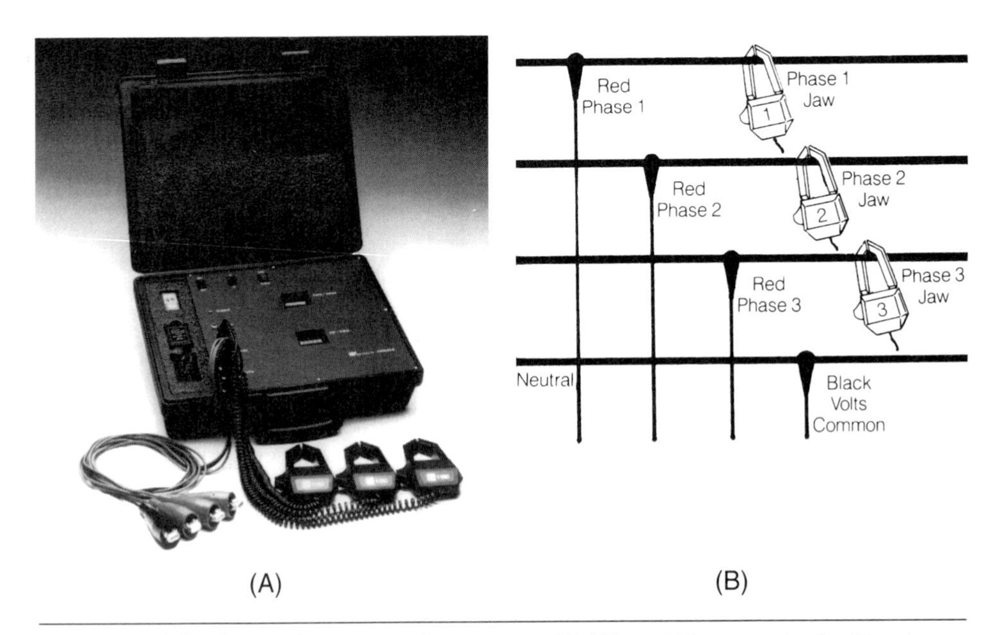

Figure 8-3 (A) Kilowatt hour meter/wattmeter. (B) Kilowatt hour meter/wattmeter used to measure power in a three-phase circuit. (Courtesy TIF Instruments, Inc.)

Figure 8-4 Simple temperature sensing device. (Courtesy TIF Instruments, Inc.)

TEMPERATURE SENSING DEVICES

The two types of temperature sensing devices in common use are the single-probe and multi-probe types.

Single-Probe Temperature Sensing Devices

A simple temperature sensing probe (shown in Figure 8-4) is designed to test either air temperatures or component temperatures. A meter of this type often has a wide range, and is referred to as a *temperature/pyrometer unit*.

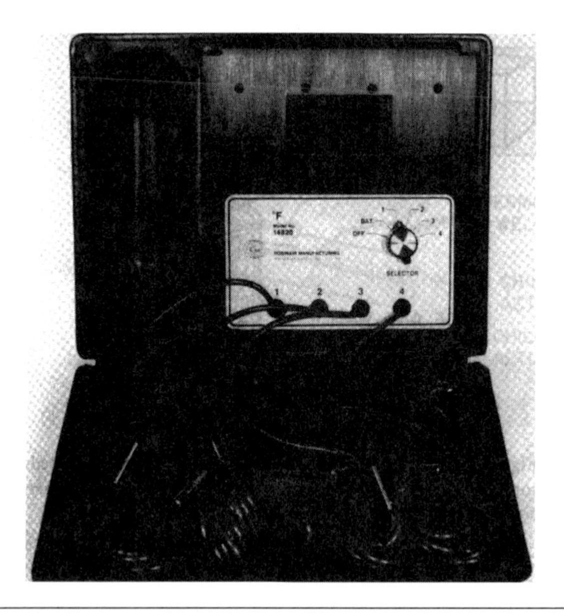

Figure 8-5 Multi-probe temperature sensing device. (By Bill Johnson)

Multi-Probe Temperature Sensing Devices

When testing for temperature drop (or rise) across a coil or testing for superheat on a refrigeration system, a multi-probe electronic thermometer can be used (Figure 8-5). This device provides simultaneous temperature readings in multiple locations without moving the probe.

CAPACITOR ANALYZERS

While an ohmmeter or the ohmmeter portion of a VOM can be used to test capacitors up to a certain point, a capacitor analyzer is a more effective way of testing a capacitor (Figure 8-6). This instrument actually applies power to the capacitor, and can also safely discharge the capacitor once the testing is complete.

TEST CORD DEVICES

There are times when troubleshooting a piece of equipment can be simplified by bypassing circuitry and start devices, and connecting power directly to a motor or compressor. Simple test cord devices are available that allow direct connection to various types of circuits and motors (Figure 8-7).

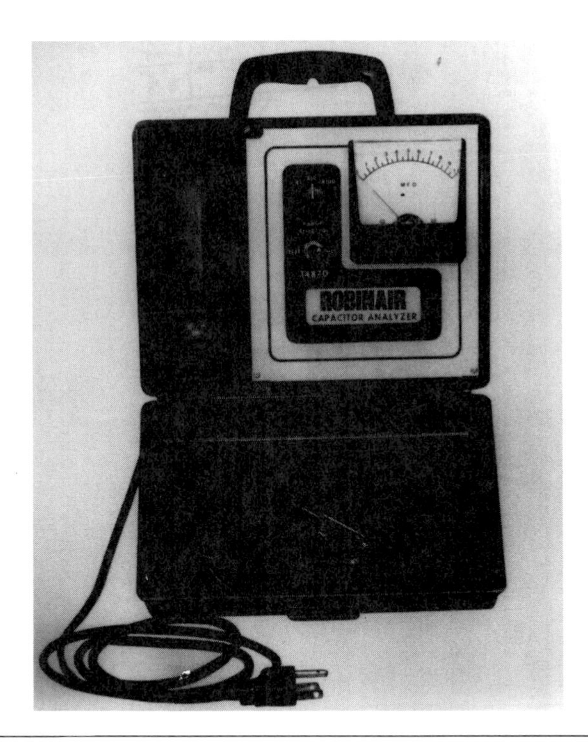

Figure 8-6 Capacitor analyzer. (Courtesy Robinair Division, SPX Corporation)

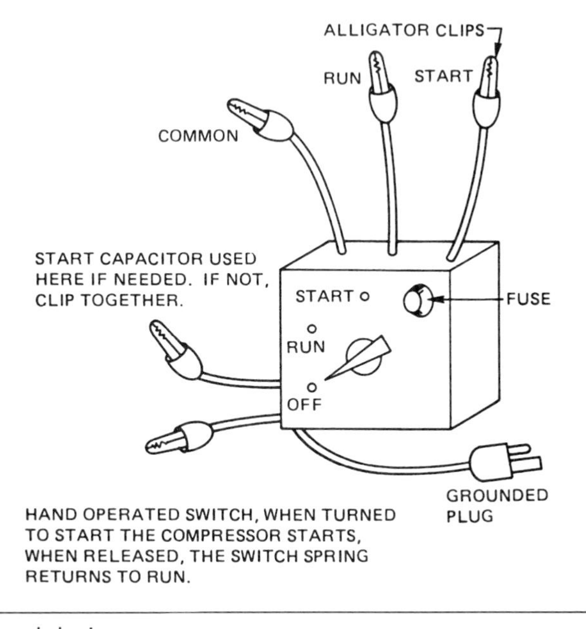

Figure 8-7 Test cord device.

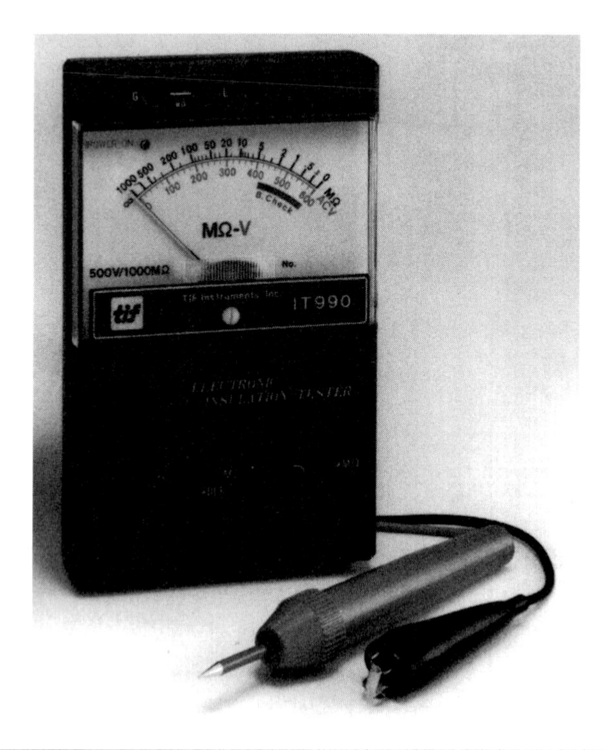

Figure 8-8 Insulation tester. (Courtesy TIF Instruments, Inc.)

Test cords also provide the option of either using a start capacitor or not using one. In addition, split cords make it easy to test a motor for proper amperage draw once it has been isolated from the equipment circuitry.

INSULATION TESTERS

An insulation tester, sometimes referred to as a *megohmmeter*, is used to perform resistance tests on motor windings (Figure 8-8). This type of meter is more efficient at testing motor windings because it supplies a higher voltage through the winding than the standard ohmmeter.

POWER FACTOR METERS

A power factor meter (Figure 8-9) is sometimes used to test electrical circuits to determine whether or not the system is operating at its highest possible efficiency.

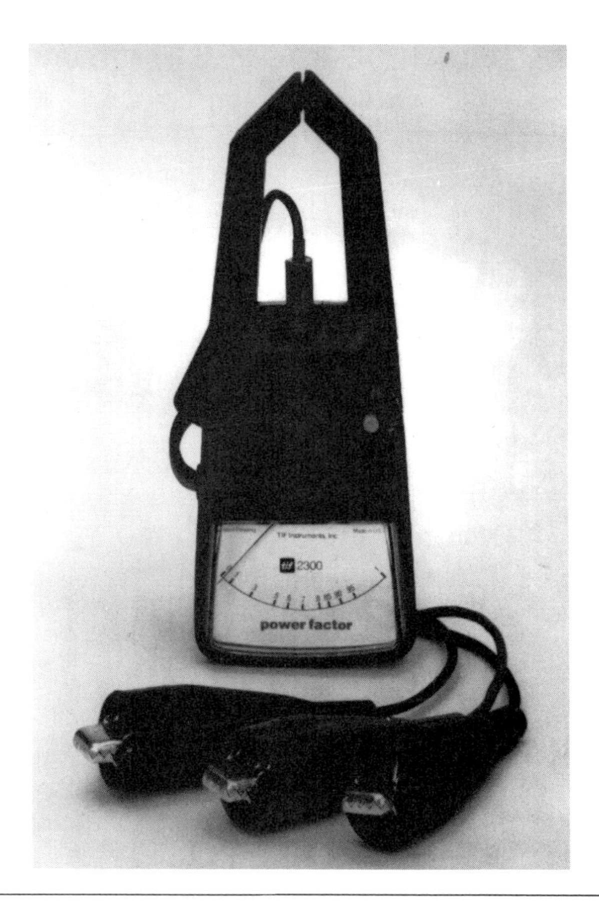

Figure 8-9 Power factor meter. (Courtesy TIF Instruments, Inc.)

USING A VOM

One way to become familiar with the operation of a VOM is to use it to test a known quantity, such as a standard wall receptacle. When using your meter to test a 120-volt wall receptacle, look closely at the position of the needle and try changing scales to a much higher voltage, just to see how the meter reads on that scale.

TROUBLESHOOTING HINT

It is always a good idea to begin by using the highest voltage scale on your meter when testing a circuit. This avoids damage to the meter caused by connecting it to a voltage higher than it has the capacity to read on a lower scale.

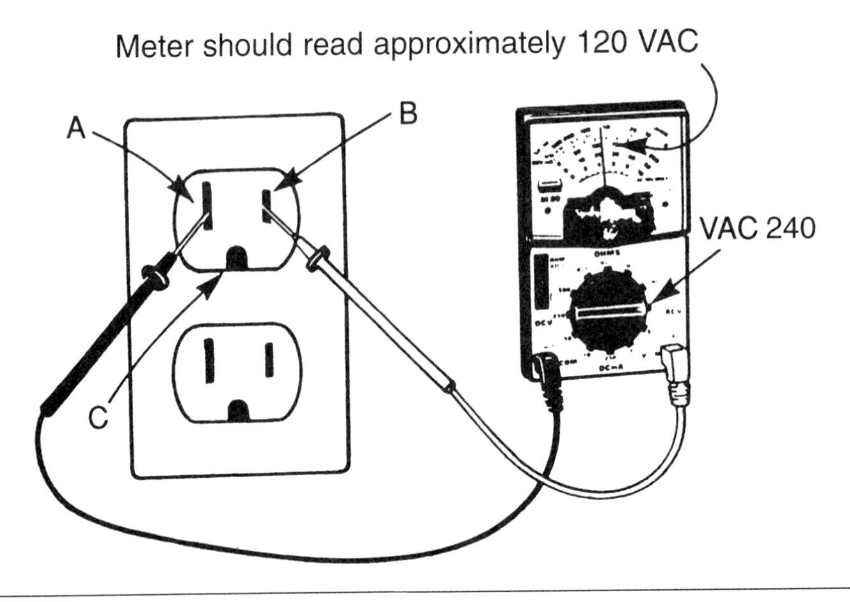

Figure 8-10 Using a VOM to test a standard wall outlet. (Note that the meter is set on the 240 VAC scale for the initial test.)

The simple test shown in Figure 8-10 can be performed to determine if a receptacle is good and if it is wired correctly. The voltage applied to the receptacle will be read between points A and B (hot to neutral) and between points B and C (hot to ground). If you do not read voltage at the points described, either the receptacle is faulty or the wiring is incorrect.

Figure 8-11 shows another simple voltage test. In this case, we are testing a circuit breaker in a panel. The voltmeter can determine if a circuit breaker is in the ON position or if it has tripped. In our illustration, the meter on the right is testing a circuit breaker that is in the closed position. The voltage is read by placing one lead of the meter on the wire connection from the single pole breaker, and the other lead on the neutral bus bar. The VOM shows a reading of 120 VAC.

The meter shown on the left in Figure 8-11 is testing a circuit breaker that has tripped. In this case, the meter reads 0 VAC.

WARNING

Always be careful when testing circuit breakers in a distribution panel. With the cover removed, the "hot" bus bars are exposed, and turning the circuit breaker off does not cut the power. Only turning off the main breaker or removing the meter breaks the circuit to the bus bars.

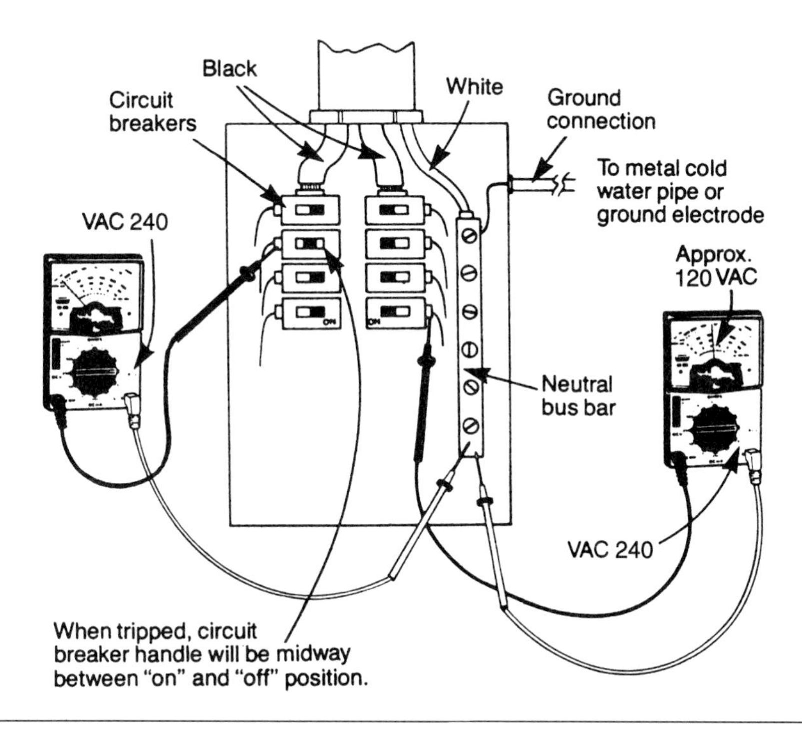

Figure 8-11 Using a VOM to test a circuit breaker.

Another important function of the standard multimeter is its ability to test for continuity. To test for continuity, you must first "zero" the meter by setting it to the lowest setting (R x 1) on the ohm scale. Then the meter can be adjusted by touching the leads together as shown in Figure 8-12.

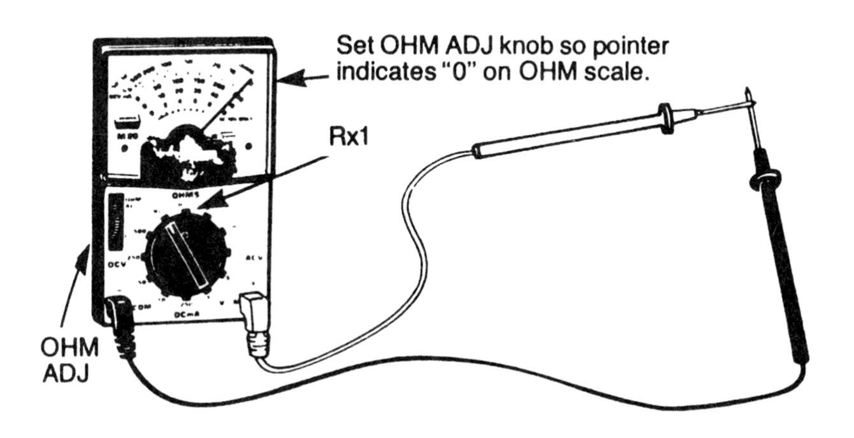

Figure 8-12 Zeroing a VOM.

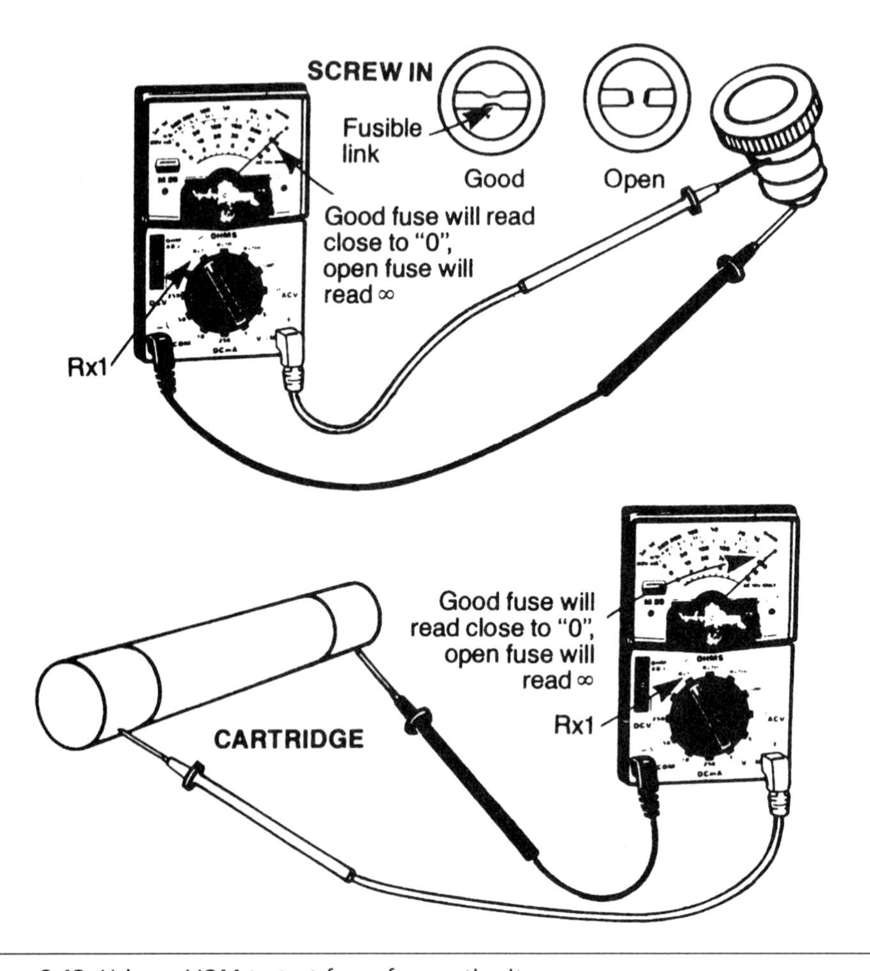

Figure 8-13 Using a VOM to test fuses for continuity.

With the leads of the meter touching, adjust the knob until the meter reading is at zero. With this adjustment accomplished, a fuse can be tested to find out if it is good or if it has blown.

Figure 8-13 shows a continuity test being performed on two types of fuses, a plug fuse that screws into a socket, and a cartridge fuse that fits in a fuse holder in a disconnect box.

TROUBLESHOOTING HINT

When testing a suspected blown cartridge fuse, do not set it down on a metal surface, such as the cabinet of an air conditioning unit. A false continuity reading will result.

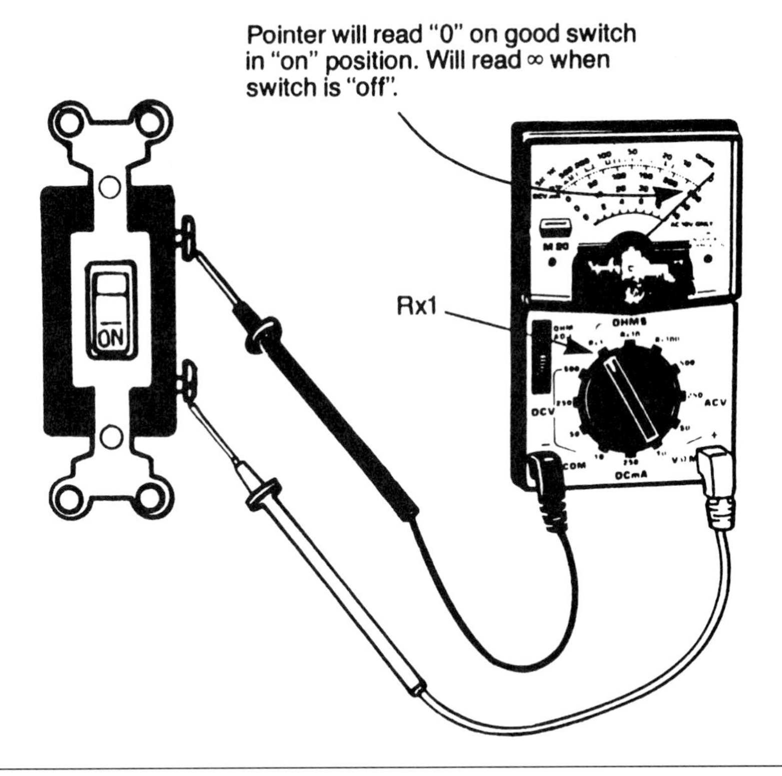

Figure 8-14 Using a VOM to test a simple switch.

With the meter in the continuity mode, you can also use it to test any type of switch to tell if it is operational. Figure 8-14 shows this test on the simplest of all switches, a single pole, single throw light switch. By placing the leads of the meter on the connections of the switch and turning it to the ON position, you can determine if the contacts inside the switch are closing. If the switch is good, the needle will rise and the meter will read zero ohms. If the switch is faulty, the needle will not rise and the meter will read infinity (∞).

When testing the switch in the OFF position, the meter should read infinity. If it shows zero resistance in this position, it is faulty and must be replaced.

UNIT EIGHT SUMMARY

In order to function effectively as an HVAC-R technician, you must be familiar with the various types of test equipment used to evaluate electrical components. Most of your workload will involve troubleshooting and evaluating electrical circuits.

A multimeter is a test instrument used to measure the voltage in a circuit, the resistance of a component, such as a coil or heating element, or the continuity across switch contacts and in a wiring harness.

An ammeter is used to check for current flow in an electrical circuit. By clamping the jaws around one wire in a circuit, you can determine the amperage draw in a circuit or of a given component.

A kilowatt/hour meter is an energy management tool used to determine energy efficiency. It can be used to obtain accumulated power consumption information over a period of time and it is helpful when trying to isolate whether or not a specific piece of equipment is to blame for excessive electric bills.

Temperature sensing devices are available in both single-probe and multi-probe types. Multi-probe temperature sensors provide an advantage in that they can measure the temperature in more than one location at a time. They are used to determine the temperature drop across an evaporator coil, the temperature rise across a condenser coil, and to test the overall efficiency of a refrigeration system.

Capacitor analyzers apply power to a capacitor for testing purposes and discharge the capacitor once the test has been completed.

Test cords are often used to eliminate start devices from HVAC-R equipment and apply power directly to compressors or other components.

Other types of test equipment include insulation testers and power factor meters. Insulation testers are used to perform resistance tests on motor windings. Power factor meters test a unit's operating efficiency.

SECTION

4

HVAC-R EQUIPMENT

This section describes various types of refrigeration and air conditioning equipment, and covers twenty specific troubleshooting problems. It begins by discussing small refrigeration systems in Unit 9, followed by commercial refrigeration equipment in Unit 10, and comfort cooling and heating equipment in Unit 11. Each unit includes several troubleshooting problems.

The first step in effective troubleshooting is fault isolation. Isolate the problem by asking yourself the following question: "Is the problem in the electrical system, refrigeration system, or air system?" Once you have identified the major inoperative system, you can begin to eliminate the possibilities.

For example, if you have a compressor that tries to start, but does not run, you can solve the problem by systematically eliminating the possibilities. They include:

1. Insufficient voltage applied to the compressor.
2. A faulty start relay.
3. A faulty start capacitor (if so equipped).
4. A faulty run capacitor (if a PSC motor).
5. The compressor itself.

How do you decide where to start and in which direction to go? One simple answer would be to check the easiest item first (the voltage applied to the circuit), then the next step could be to "hopscotch" directly to possibility number 5. Using a test cord to test the compressor directly is an effective way to eliminate it as a possibility. Depending on the results of this test, you would either have diagnosed the problem or found it necessary to jump back to possibility number 2. This systematic troubleshooting approach, along with the process of elimination, will allow you to quickly solve most of the problems you will encounter as an HVAC-R technician.

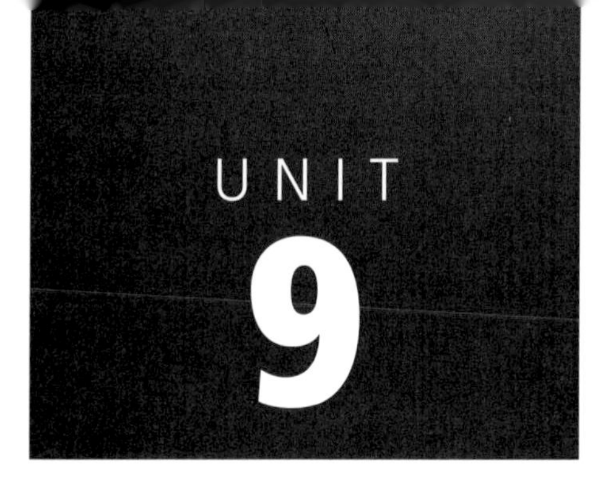

Small Refrigeration Systems

APPLICATIONS

Small refrigeration systems are used in domestic refrigerators and freezers, beverage dispensing machines, reach-in refrigerators used in restaurants, room (window) air conditioners, and small drinking fountains and water coolers.

Small refrigeration systems use fractional horsepower compressors, and with the exception of water coolers that use an automatic expansion valve, most use a capillary tube as the metering device. Figure 9-1 shows a basic illustration of a small refrigeration system.

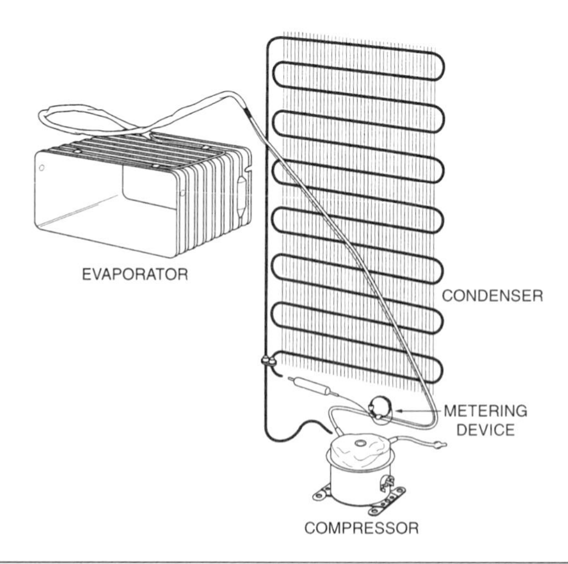

Figure 9-1 An illustration of the four basic components as they are used in a conventional refrigerator.

This illustration of a conventional refrigerator shows the four basic components of the refrigeration system: the compressor, condenser, evaporator, and metering device.

In this system (and most small refrigeration systems), the metering device and capillary tube are bonded together to form what is known as the *heat exchanger*. (Refer to Figure 9-2.)

Bonding the capillary tube to the suction line serves two purposes. It helps to prevent any liquid refrigerant from getting to the compressor, and also helps to assure that the refrigerant will enter the evaporator at the right temperature.

This system works because the capillary tube is warm and will give up some of its heat to the cooler suction line. And, if any liquid **was** traveling down the suction line to the compressor, the heat from the capillary tube would cause it to change state from a liquid to a vapor.

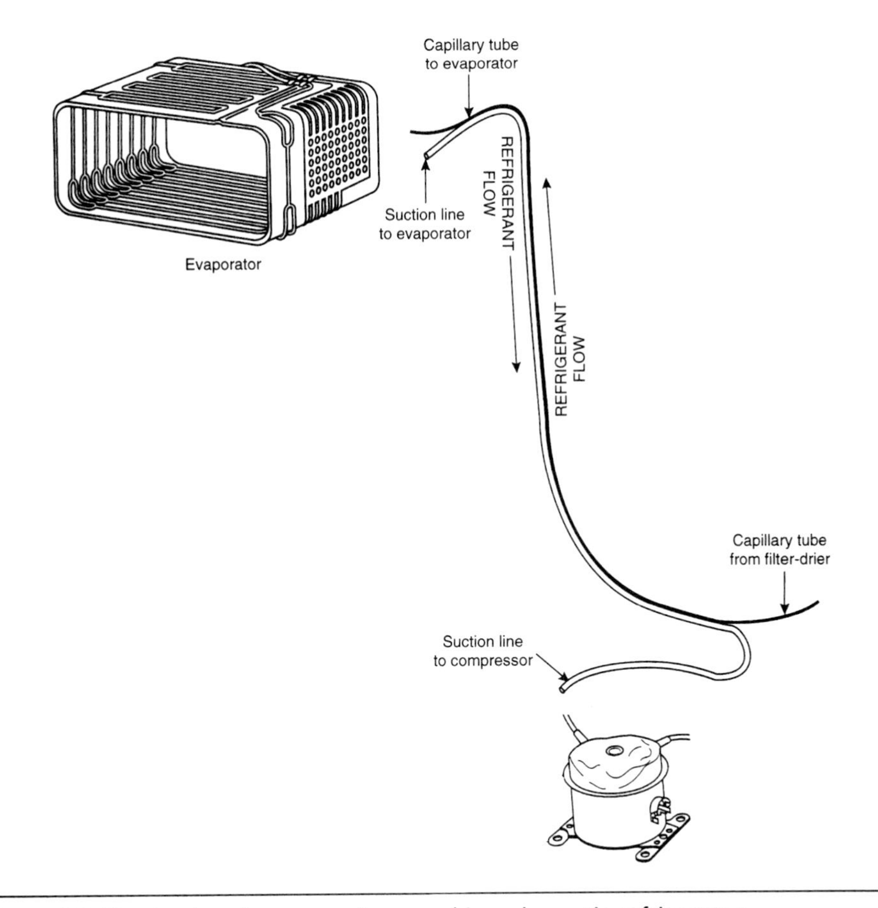

Figure 9-2 The heat exchanger system used in a domestic refrigerator.

TROUBLESHOOTING HINT

If you are working with this type of system and it is operating inefficiently, check to see if the capillary tube has separated from the suction line. If there is not good metal-to-metal contact between the two tubes, the system can lose up to 15% of its refrigeration capacity.

To repair this problem, thermal contact can be re-established by sanding the capillary tube and suction line to eliminate any corrosion buildup. After sanding, use a thermal mastic material that will create a good thermal bond between the two components when it hardens.

Small refrigeration systems can also utilize forced air in both the condenser and evaporator sides of the system. In forced-air units that operate with a low temperature coil, some method of defrost is used to clear the frost from the evaporator several times each day. Defrost is usually accomplished with an electric heater, although hot gas defrost may be found in some reach-in refrigerators in restaurants.

Another application of the small refrigeration system is the room air conditioner (Figure 9-3). These units also use a capillary tube as the metering device, but do not use the heat exchanger system. In some applications, you may find that multiple capillary tubes are used.

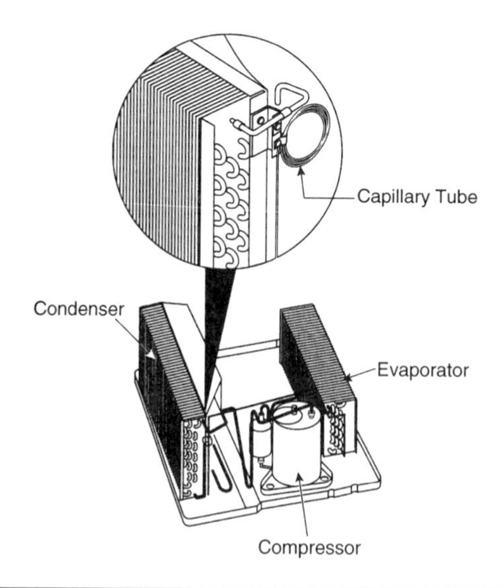

Figure 9-3 The capillary tube is also used as a metering device in window air conditioners.

SERVICE PROCEDURES

Small refrigeration systems rarely include factory-installed access valves, except in situations such as some fractional horsepower commercial equipment found in restaurants. This means that if you need to service the sealed system, an access valve must be installed.

Two types of valves used for installation on small refrigeration systems are the tap-line valve and the permanent access valve. The tap-line valve is a saddle-type device that is bolted around the tubing, then pierces the copper line of the system. This allows you to access the sealed system without installing a valve that requires soldering. The permanent access valve, which is also a piercing valve, is considered permanent because, unlike the tap-line valve, it is brazed to the refrigerant line. The refrigerant must be removed from the system in order to install this type of valve. Figure 9-4 shows a tap-line valve, and Figure 9-5 shows the procedure for installing a permanent access valve.

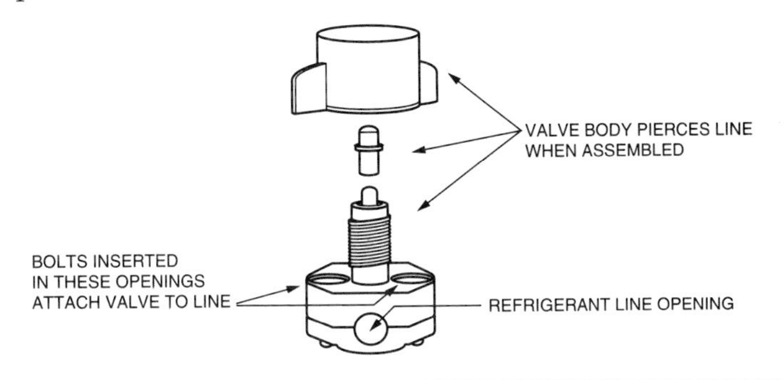

Figure 9-4 A saddle-type access valve.

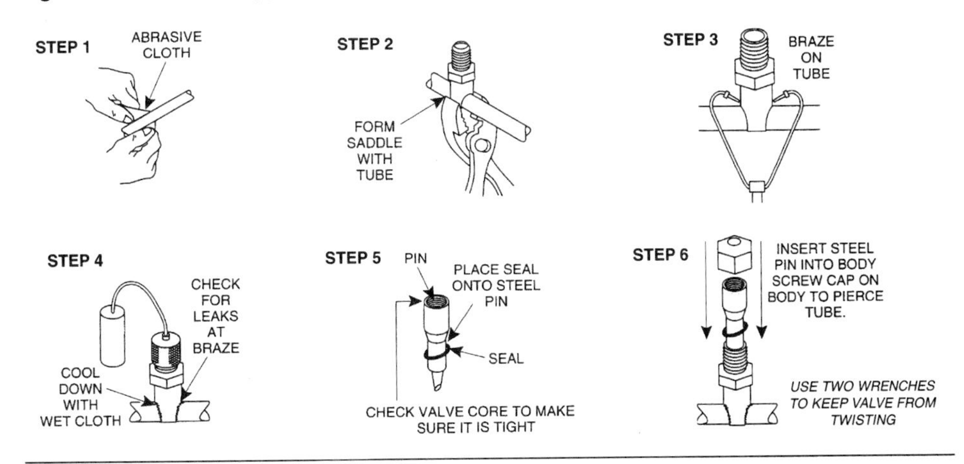

Figure 9-5 The proper procedure for installing a permanent access valve.

TEMPERATURE °F	REFRIGERANT				TEMPERATURE °F	REFRIGERANT				TEMPERATURE °F	REFRIGERANT			
	12	22	134a	502		12	22	134a	502		12	22	134a	502
−60	19.0	12.0		7.2	12	**15.8**	**34.7**	**13.2**	**43.2**	42	**38.8**	**71.4**	**37.0**	**83.8**
−55	17.3	9.2		3.8	13	**16.4**	**35.7**	**13.8**	**44.3**	43	**39.8**	**73.0**	**38.0**	**85.4**
−50	15.4	6.2		0.2	14	**17.1**	**36.7**	**14.4**	**45.4**	44	**40.7**	**74.5**	**39.0**	**87.0**
−45	13.3	2.7		**1.9**	15	**17.7**	**37.7**	**15.1**	**46.5**	45	**41.7**	**76.0**	**40.1**	**88.7**
−40	11.0	**0.5**	14.7	**4.1**	16	**18.4**	**38.7**	**15.7**	**47.7**	46	**42.6**	**77.6**	**41.1**	**90.4**
−35	8.4	**2.6**	12.4	**6.5**	17	**19.0**	**39.8**	**16.4**	**48.8**	47	**43.6**	**79.2**	**42.2**	**92.1**
−30	5.5	**4.9**	9.7	**9.2**	18	**19.7**	**40.8**	**17.1**	**50.0**	48	**44.6**	**80.8**	**43.3**	**93.9**
−25	2.3	**7.4**	6.8	**12.1**	19	**20.4**	**41.9**	**17.7**	**51.2**	49	**45.7**	**82.4**	**44.4**	**95.6**
−20	**0.6**	**10.1**	3.6	**15.3**	20	**21.0**	**43.0**	**18.4**	**52.4**	50	**46.7**	**84.0**	**45.5**	**97.4**
−18	**1.3**	**11.3**	2.2	**16.7**	21	**21.7**	**44.1**	**19.2**	**53.7**	55	**52.0**	**92.6**	**51.3**	**106.6**
−16	**2.0**	**12.5**	0.7	**18.1**	22	**22.4**	**45.3**	**19.9**	**54.9**	60	**57.7**	**101.6**	**57.3**	**116.4**
−14	**2.8**	**13.8**	**.3**	**19.5**	23	**23.2**	**46.4**	**20.6**	**56.2**	65	**63.8**	**111.2**	**64.1**	**126.7**
−12	**3.6**	**15.1**	**1.2**	**21.0**	24	**23.9**	**47.6**	**21.4**	**57.5**	70	**70.2**	**121.4**	**71.2**	**137.6**
−10	**4.5**	**16.5**	**2.0**	**22.6**	25	**24.6**	**48.8**	**22.0**	**58.8**	75	**77.0**	**132.2**	**78.7**	**149.1**
−8	**5.4**	**17.9**	**2.8**	**24.2**	26	**25.4**	**49.9**	**22.9**	**60.1**	80	**84.2**	**143.6**	**86.8**	**161.2**
−6	**6.3**	**19.3**	**3.7**	**25.8**	27	**26.1**	**51.2**	**23.7**	**61.5**	85	**91.8**	**155.7**	**95.3**	**174.0**
−4	**7.2**	**20.8**	**4.6**	**27.5**	28	**26.9**	**52.4**	**24.5**	**62.8**	90	**99.8**	**168.4**	**104.4**	**187.4**
−2	**8.2**	**22.4**	**5.5**	**29.3**	29	**27.7**	**53.6**	**25.3**	**64.2**	95	**108.2**	**181.8**	**114.0**	**201.4**
0	**9.2**	**24.0**	**6.5**	**31.1**	30	**28.4**	**54.9**	**26.1**	**65.6**	100	**117.2**	**195.9**	**124.2**	**216.2**
1	**9.7**	**24.8**	**7.0**	**32.0**	31	**29.2**	**56.2**	**26.9**	**67.0**	105	**126.6**	**210.8**	**135.0**	**231.7**
2	**10.2**	**25.6**	**7.5**	**32.9**	32	**30.1**	**57.5**	**27.8**	**68.4**	110	**136.4**	**226.4**	**146.4**	**247.9**
3	**10.7**	**26.4**	**8.0**	**33.9**	33	**30.9**	**58.8**	**28.7**	**69.9**	115	**146.8**	**242.7**	**158.5**	**264.9**
4	**11.2**	**27.3**	**8.6**	**34.9**	34	**31.7**	**60.1**	**29.5**	**71.3**	120	**157.6**	**259.9**	**171.2**	**282.7**
5	**11.8**	**28.2**	**9.1**	**35.8**	35	**32.6**	**61.5**	**30.4**	**72.8**	125	**169.1**	**277.9**	**184.6**	**301.4**
6	**12.3**	**29.1**	**9.7**	**36.8**	36	**33.4**	**62.8**	**31.3**	**74.3**	130	**181.0**	**296.8**	**198.7**	**320.8**
7	**12.9**	**30.0**	**10.2**	**37.9**	37	**34.3**	**64.2**	**32.2**	**75.8**	135	**193.5**	**316.6**	**213.5**	**341.2**
8	**13.5**	**30.9**	**10.8**	**38.9**	38	**35.2**	**65.6**	**33.2**	**77.4**	140	**206.6**	**337.2**	**229.1**	**362.6**
9	**14.0**	**31.8**	**11.4**	**39.9**	39	**36.1**	**67.1**	**34.1**	**79.0**	145	**220.3**	**358.9**	**245.5**	**385.0**
10	**14.6**	**32.8**	**11.9**	**41.0**	40	**37.0**	**68.5**	**35.1**	**80.5**	150	**234.6**	**381.5**	**262.7**	**408.4**
11	**15.2**	**33.7**	**12.5**	**42.1**	41	**37.9**	**70.0**	**36.0**	**82.1**	155	**249.5**	**405.1**	**280.7**	**432.9**

Note: Bold figures in the refrigerant columns represent gauge pressure; remaining figures in those columns represent vacuum pressure.

Figure 9-6 Temperature/pressure chart displaying pressure in psig and in. Hg vacuum.

TROUBLESHOOTING THE REFRIGERATION SYSTEM

The first step in troubleshooting a suspected sealed system problem is to establish what is *supposed to be happening* in a system that is operating normally. One way to accomplish this is to use a temperature/pressure (T/P) chart along with your gauges. T/P charts show the pressure of a refrigerant at a given temperature (Figure 9-6).

You will recall from Unit 3 that as the temperature of a refrigerant goes up, the pressure also goes up, and as the temperature goes down, the pressure also goes down.

To decide if a unit is operating normally, you must know something about its design and operating conditions. Consider the following hypothetical situation:

- The system is using refrigerant HFC-134a.
- The evaporator coil design temperature is -10°F.
- The ambient temperature is 75°F.

With the above facts in hand, you can determine the pressures you should read on a normally operating system.

To use the T/P chart and your gauges to evaluate the hypothetical refrigeration system, follow these steps:

STEP ONE: Read down the temperature column at the left to find the evaporator coil design temperature (-10°F).

STEP TWO: Read across to the appropriate refrigerant column (134a) to find the low side operating pressure of this unit (2 psig).

STEP THREE: Find the ambient temperature (75°F) in the temperature column.

STEP FOUR: Add 30°F to this figure to compensate for the heat of compression in the system.

STEP FIVE: Read across from the adjusted temperature (105°F) to the correct refrigerant column (134a) to find that the high side operating pressure of this unit should be 135 psig.

With these steps accomplished, we now know that our system should have a low side operating pressure of 2 psig, and a high side operating pressure of 135 psig. If we read pressures other than these, it can be established that there is either a problem in the sealed system, or system performance is being affected by another factor, such as improper air flow over a coil (something that should have been eliminated as a possibility before a tap-line valve was installed on the system.)

With other possibilities such as the electrical and air flow systems eliminated, the information established on a properly functioning sealed system can be applied to units that are not operating properly. The following problems are commonly found in sealed systems:

- Overcharged system.
- Undercharged system with a low side leak.
- Completely restricted system.
- System with an inefficient compressor.

Keep in mind that to be an effective troubleshooter, you must use all the information at your disposal. This includes visual inspection, listening, and using your sense of touch, along with using gauges to record the operating pressures in the system, and an ammeter to evaluate the compressor's electrical operation. To illustrate this process, systems with the problems listed above are shown in Figures 9-7 through 9-10.

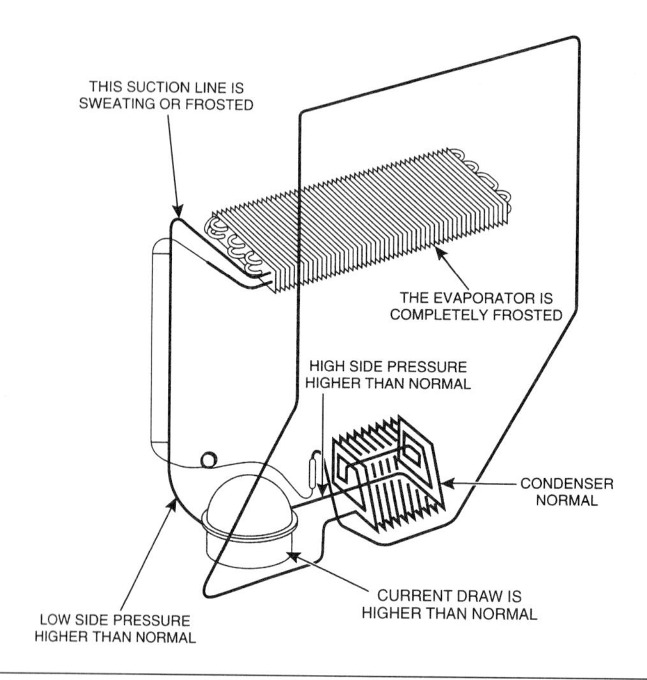

Figure 9-7 An overcharged sealed system in a refrigerator.

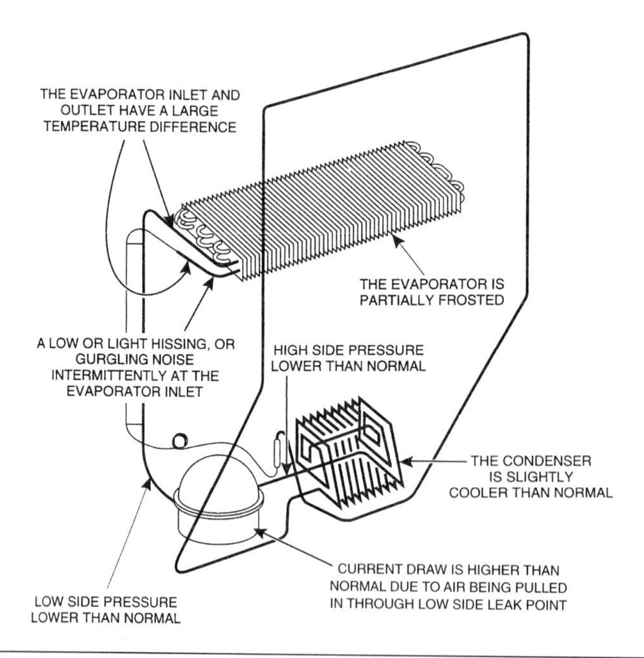

Figure 9-8 An undercharged sealed system in a refrigerator.

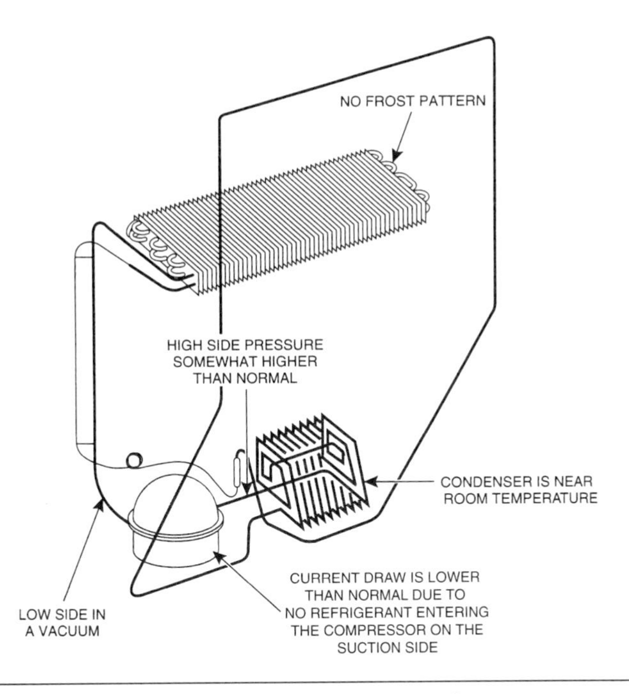

Figure 9-9 A completely restricted sealed system in a refrigerator.

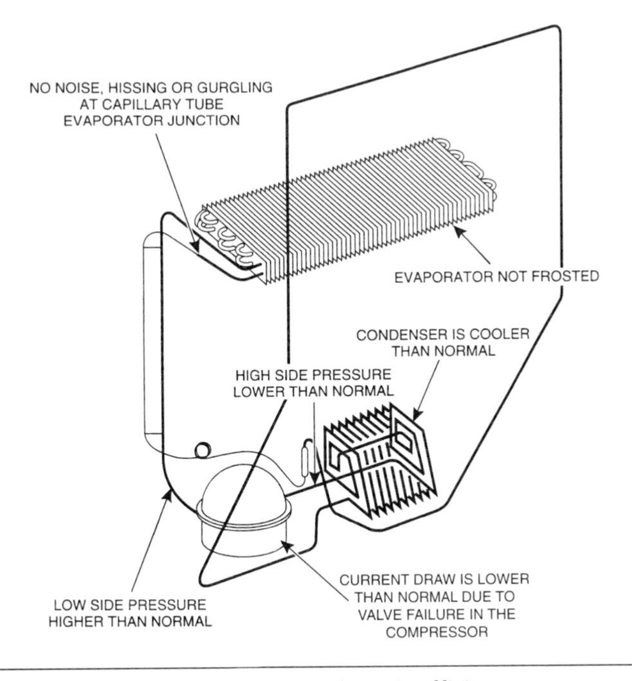

Figure 9-10 A refrigerator sealed system with an inefficient compressor.

ELECTRICAL SYSTEM SERVICE PROCEDURES

When troubleshooting the electrical section on small refrigeration systems, you must become familiar with the two types of wiring diagrams used, the *schematic* and the *pictorial*.

The schematic diagram, such as the one shown in Figure 9-11, shows the electrical circuitry of the unit. The pictorial diagram (see Figure 9-12) shows the location of the components and the actual routing of the wiring harness.

The three basic electrical service procedures followed when tracking down a problem in an electrical system involve the use of meters that measure voltage, resistance, and amperage draw.

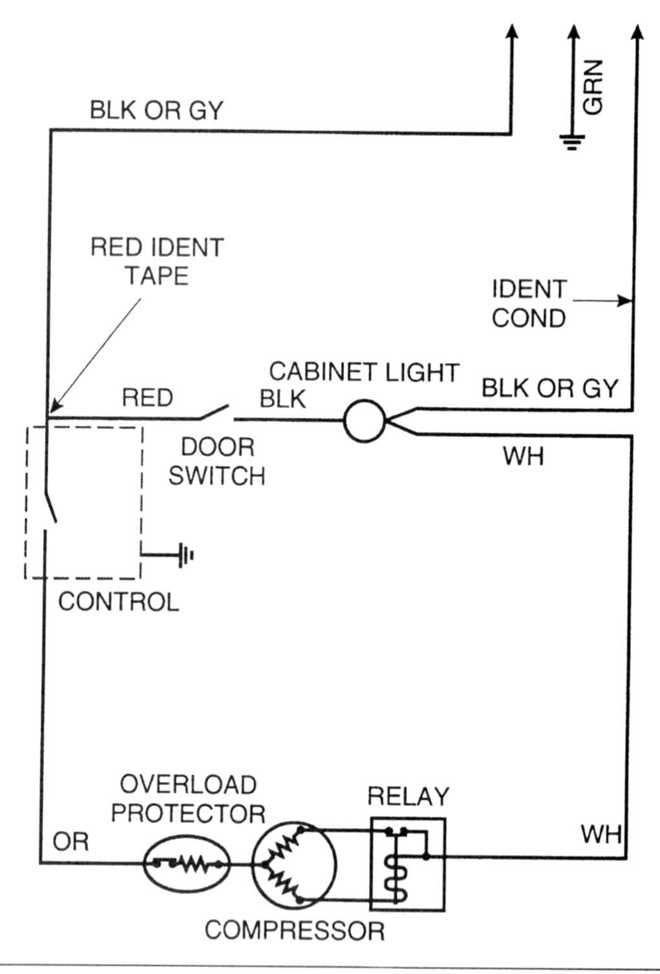

Figure 9-11 A simple schematic diagram for a conventional refrigerator.

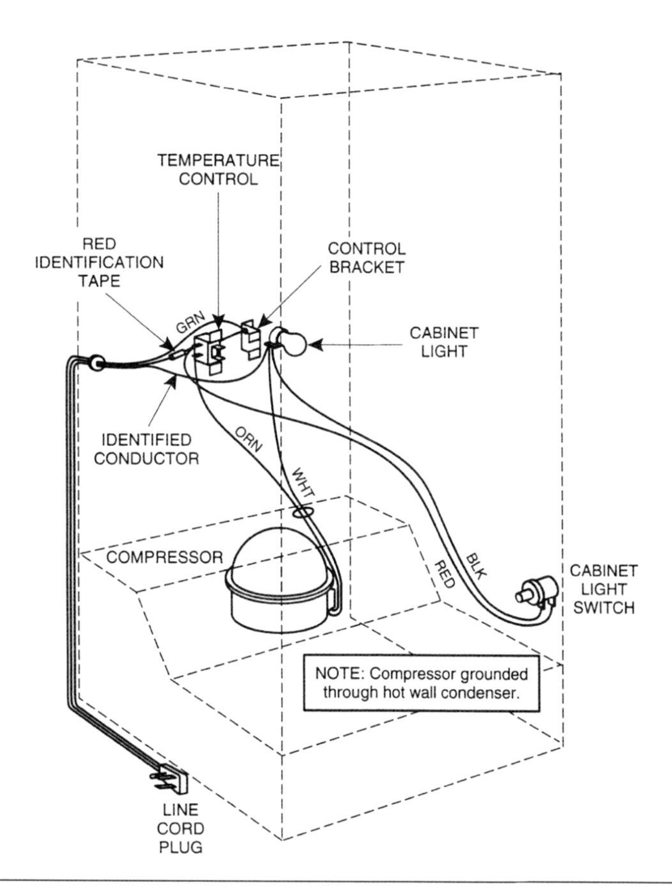

Figure 9-12 A pictorial diagram showing the electrical system of a conventional refrigerator.

WARNING

When using an ohmmeter to test components or a wire, the power should **always** be disconnected, and in order to effectively test a component, it will be necessary to remove at least one wire to prevent a false reading due to the wiring harness itself or other components in the electrical system.

When using a voltmeter, you are working with a "hot" system and must use extreme care to prevent electrical shock or damage that would occur if a lead from your meter touched a live wire at the same time it touched ground. Also, always exercise caution when using an ammeter and working with wires that are carrying current.

In the three troubleshooting exercises that follow, step-by-step procedures are outlined for the use of test meters and gauges.

PROBLEM 1 *UNIT NOT RUNNING*

In this 120-volt circuit, shown in the schematic in Figure 9-11, the "hot" wire is shown as black or gray (BLK or GY) and the neutral wire is shown as white (WH). The hot wire becomes orange (OR) after it leaves the control, which is wired in series with the overload protector and common terminal of the compressor.

The steps to troubleshoot this problem are as follows:

STEP ONE: Test the wall receptacle for 120 volts. This test shows a positive result and voltage is read. Proceed to step two.

STEP TWO: Test for voltage at the overload and relay assembly. (See Figure 9-13.) This test also shows positive results. Proceed to step three.

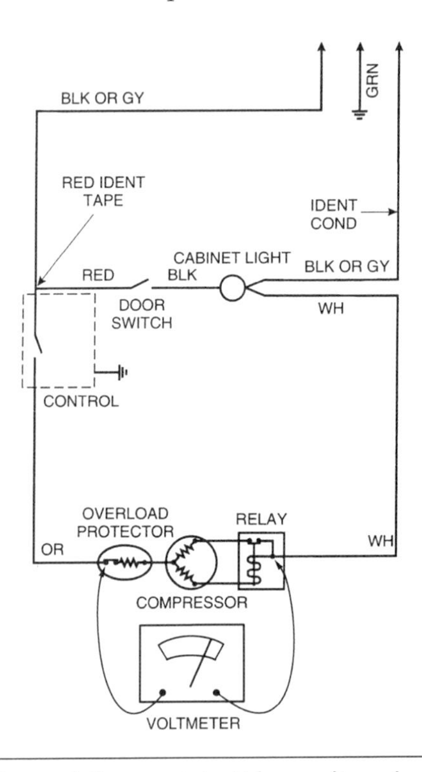

Figure 9-13 Test for voltage at these points. When voltage is confirmed, continue to test the circuit.

STEP THREE: Since the initial voltage test at the overload and relay assembly showed positive, there is no need to go "backwards" and test the control, because in this case, it is doing its assigned job, which is to complete a circuit to the overload via the orange wire. The next step is to check for voltage, as shown in Figure 9-14.

With this test, we will assume that no voltage is indicated by the meter. What is the solution to this problem?

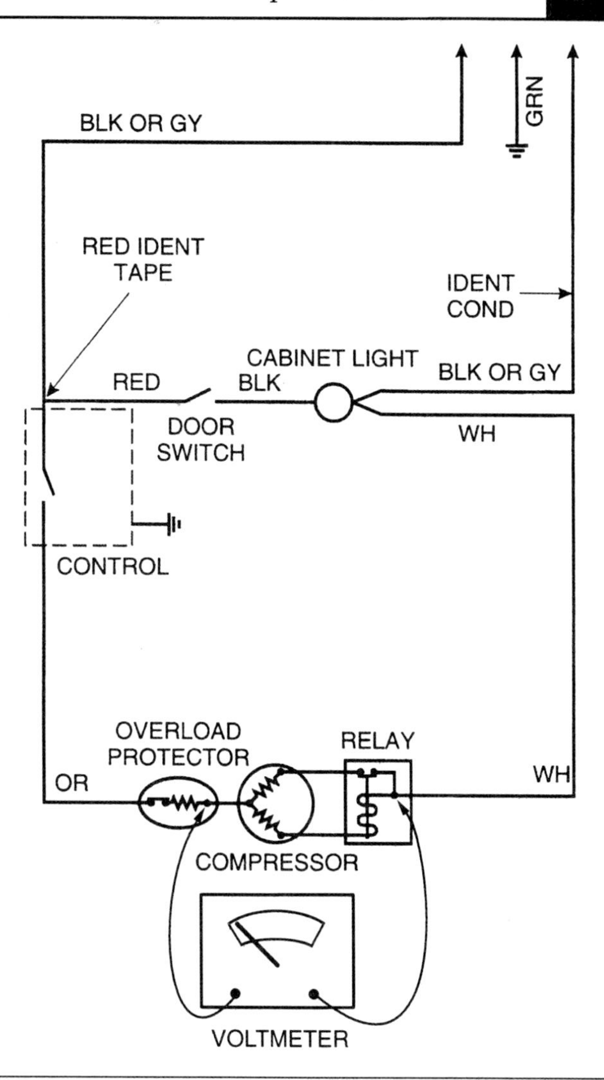

Figure 9-14 Continuing the circuit test on a "no run" complaint. No voltage is read at this point.

PROBLEM 2 COMPRESSOR TRIES TO START BUT DOES NOT RUN

With this problem, there is no need to test for voltage at the receptacle, nor to test for voltage at the orange and white wires.

STEP ONE: Disconnect the relay and overload protector and connect a test cord assembly to the common, start, and run terminals of the compressor. (See Figure 9-15.) Use the test cord to operate the compressor independently from the equipment wiring. The compressor starts and runs. Proceed to step two.

STEP TWO: Use an ammeter to test the compressor for proper amperage draw.

The ammeter test shows that the compressor is drawing the correct amount of current. Which component is at fault and must be replaced?

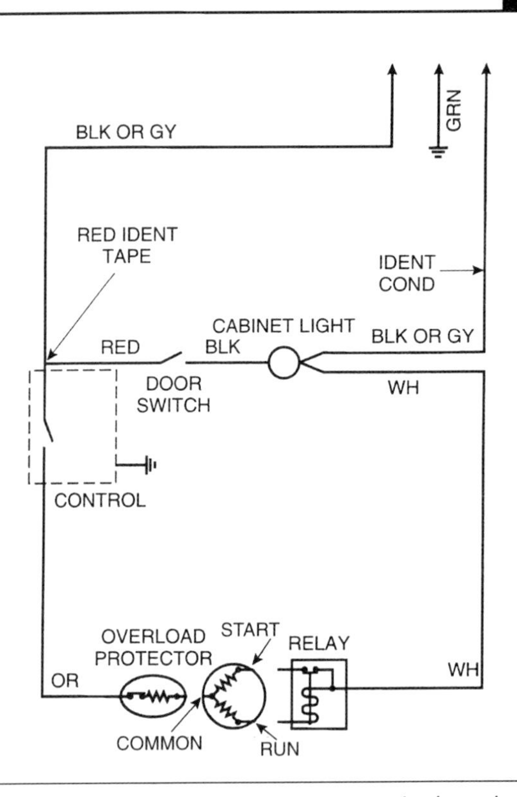

Figure 9-15 Connect a test cord to the compressor terminals and test the running compressor amperage draw.

PROBLEM 3 *UNIT IS RUNNING BUT IS NOT COOLING PROPERLY*

In this problem, all electrical and air flow problems have been eliminated, and the diagnosis must be accomplished by testing the sealed system. This system uses HFC-134a and has an evaporator coil design temperature of -10°F.

STEP ONE: Install tap-line valves on the suction line to gain access to the low pressure side of the system, and on the compressor discharge line to gain access to the high pressure side of the system. (See Figure 9-16.) Proceed to step two.

STEP TWO: Connect the low side charging hose from your gauge set to the suction line and install the high side charging hose to the discharge line access valve. Proceed to step three.

STEP THREE: Record the ambient temperature. (In this case, 75°F.) Proceed to step four.

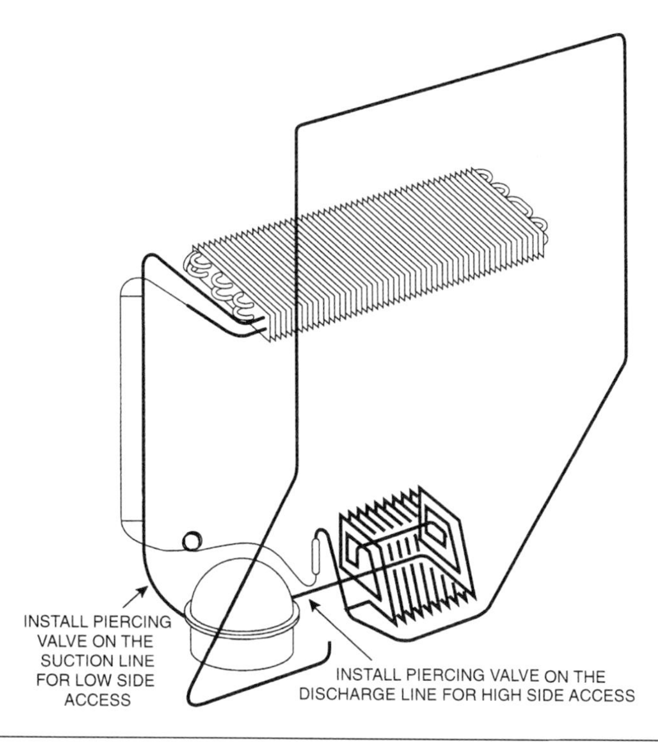

Figure 9-16 The proper installation points for tap-line valves.

STEP FOUR: Compare the pressure readings from the operating system to the pressures calculated from the temperature/pressure chart.

The readings from the system are 15 psig on the low side and 155 psig on the high side, as shown in Figure 9-17. What is the problem that must be corrected?

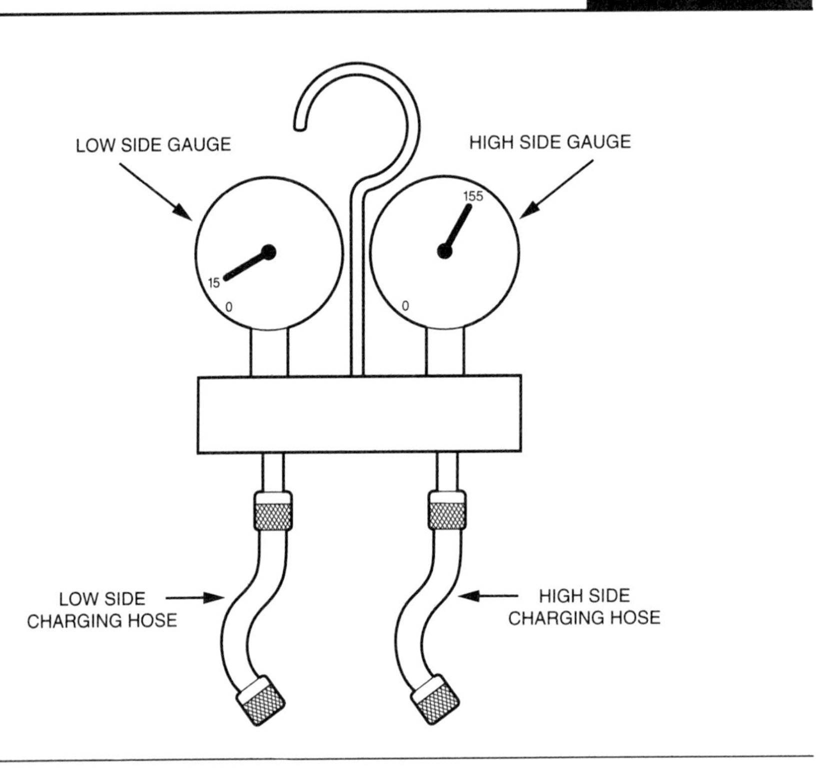

Figure 9-17 Sample refrigeration system readings.

UNIT NINE SUMMARY

The most common metering device found on small refrigeration systems is the capillary tube. In many cases, the capillary tube and suction line are soldered together to form a heat exchanger. In the event of separation of the heat exchanger, the efficiency of the refrigeration system can be affected by up to 15%.

Small refrigeration systems use both forced-air and static condensers. Both convection and forced-air evaporators are also found in these systems. In the application of a forced-air evaporator, some method of defrost is used. The defrost method may be an electric heating element or it may be a hot gas defrost system.

When servicing small refrigeration systems, you must install access valves. Unlike HVAC or commercial refrigeration systems, domestic equipment does not come equipped with valves that allow the connection of a gauge set. Two types of valves used on domestic refrigeration systems are the tap-line valve (which is considered to be temporary) and the permanent access valve, which is soldered into place.

One way to establish normal operating conditions for a small refrigeration system is to use a temperature/pressure chart. Once the normal operating conditions are established, you can compare them to the actual operating conditions and use the information to determine the problem.

When servicing small refrigeration equipment, you must become familiar with both the schematic and pictorial wiring diagrams. The schematic diagram shows the electrical circuitry of the unit, while the pictorial shows the location of the components.

UNIT 10

Commercial Refrigeration Systems

COMMERCIAL REFRIGERATION SYSTEM APPLICATIONS

Commercial refrigeration systems include equipment used in retail establishments, such as grocery stores (Figure 10-1), convenience stores, and restaurants, as well as equipment used in large comfort cooling systems, such as office buildings and apartment complexes.

Some commercial systems use fractional horsepower compressors, but for the majority of applications, larger compressors are used. Also, unlike small refrigeration systems, the condensers in commercial applications will be either a forced-air type or water cooled.

Figure 10-1 Display cases are used to safely store food displayed for sale. Some display cases are open to allow the customer to reach in without opening a door; others have doors. Closed cases are the most energy efficient. (Courtesy Hill Phoenix, Richmond, VA)

Commercial equipment also differs from small refrigeration systems in regard to the metering device. The thermostatic expansion valve is most commonly used on commercial refrigeration equipment. In newer chiller systems used in large comfort cooling systems, a solid state control system that includes an electronic expansion valve (EEV) is used.

GROCERY STORE EQUIPMENT

Grocery stores make extensive use of refrigeration equipment for food preservation and merchandising. While the display and food storage cases are located throughout the store, the compressors are located in an equipment room at the rear of the store.

Also located in or near the equipment room is the condenser that will be common to almost all of the equipment in the store. The condenser will most often be the evaporative type, in which water is sprayed directly onto the tubing while an air handler forces air across the tubing surface (Figure 10-2). In some cases, the store may be equipped with a heat reclaim system that makes use of the heat during the winter months, using an air handling system to circulate warm air throughout the store.

In a heat reclaim system, solenoid valves are often used to redirect the hot gas to tubing located in an air handler rather than to the evaporative condenser. Figure 10-3 shows a typical heat reclaim system.

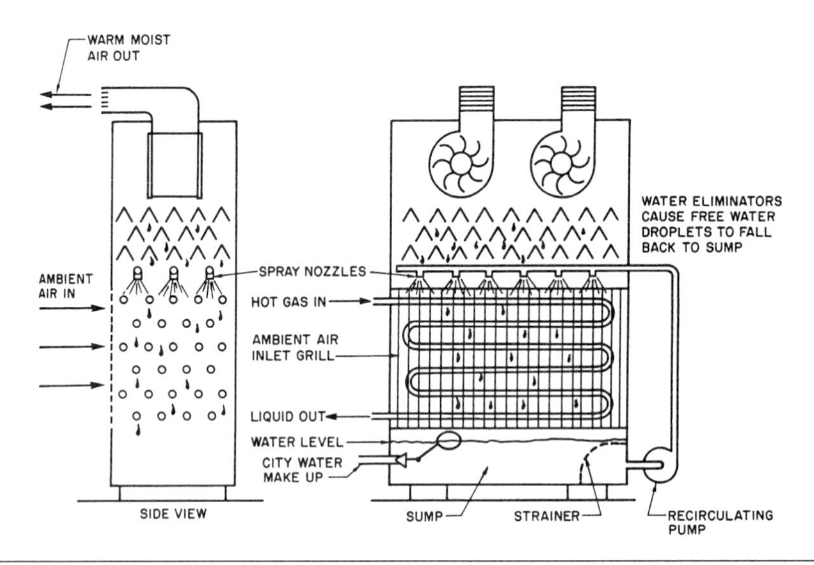

Figure 10-2 Water recirculates in the evaporative condenser. The condenser tubes are in the tower rather than in a condenser shell located in a building.

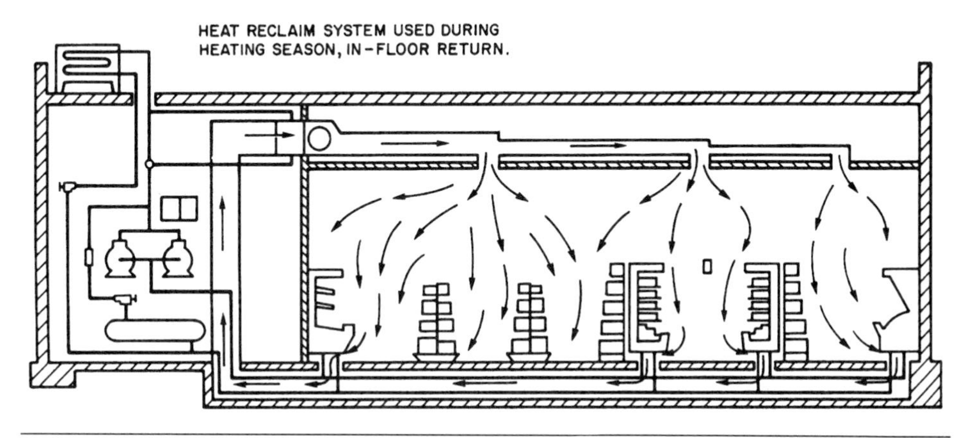

Figure 10-3 A heat reclaim system can supply heat to the store.

A heat reclaim system with alternate routing for the condenser tubing is one reason an equipment room can appear intimidating. Couple this factor with the dozen or more compressors located on the equipment room rack, a control panel for each compressor, pressure switches, oil pressure safety switches, solenoid valves, and piping to all the display and food cases, and you begin to understand the need to become familiar with your work environment before attempting any troubleshooting or repairs.

Display Cases

Two types of display cases used in grocery stores are the open display case and the closed display case. Overall, a closed display case is more efficient, but open display cases are necessary for customer convenience. Figure 10-4 shows the air flow pattern in a display case.

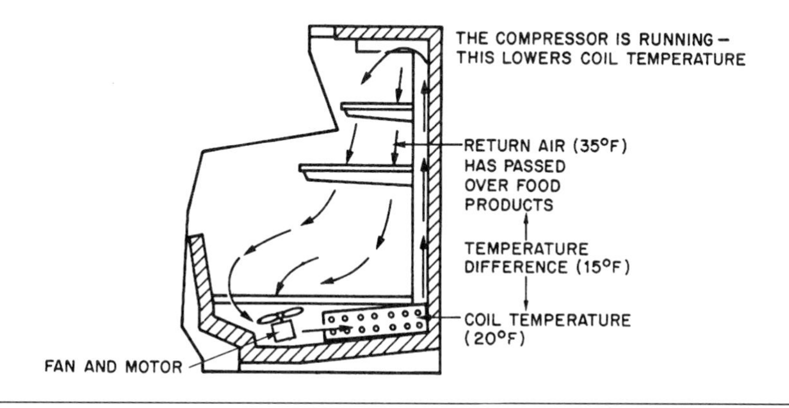

Figure 10-4 The air flow pattern in a typical open display case.

Walk-In Refrigerators

Grocery stores also make extensive use of walk-in refrigeration equipment. Low temperature walk-in freezers in which the temperature is maintained near 0°F are used to preserve ice cream and other items that are sold frozen. Higher temperature walk-in areas, such as those used for meat storage (usually near 35°F) or for produce storage (with a design temperature near 40°F) are also found in grocery stores.

TROUBLESHOOTING HINT

If you respond to a service call on a walk-in freezer, one of the easiest things to check is the operation of the evaporator fan(s). Most equipment is designed with a switch that breaks the circuit to the fan motor when the door is opened. The evaporator fan or fans will not start until the door is properly closed. *(Keep in mind that this hint applies as long as the system is in the refrigeration cycle, not the defrost cycle. Evaporator fans are traditionally shut down during the defrost mode whether the system uses hot gas or electric elements.)*

RESTAURANT EQUIPMENT

Refrigeration applications in restaurants include walk-in and reach-in refrigerators and freezers, and ice machines.

Refrigerators

Like grocery stores, restaurants also make use of walk-in refrigeration equipment in both the low temperature and higher temperature ranges. A more common piece of equipment used in restaurant kitchens is the reach-in refrigerator. Shown in Figure 10-5, it maintains a design temperature of about 35°F.

Freezers

It is also common to find a reach-in freezer in commercial kitchen applications. The cabinet design is identical to that of the reach-in refrigerator; the difference lies in the design of the refrigeration system. Reach-in freezers are capable of maintaining a design temperature of 0 to 5°F. A reach-in freezer will also use some type of defrost system to clear the evaporator coil periodically.

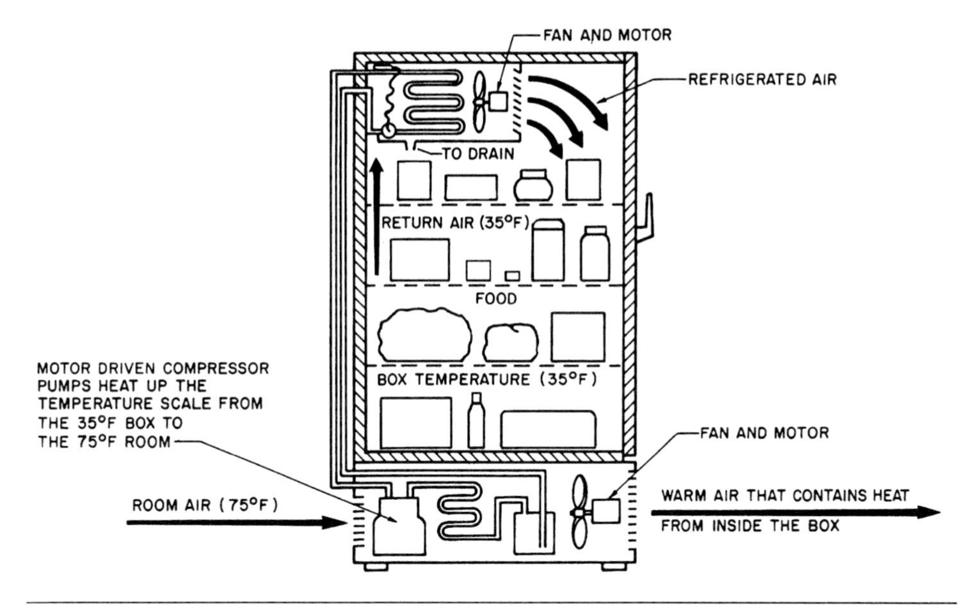

Figure 10-5 A reach-in refrigerator often found in restaurants.

Commercial Ice Machines

Commercial ice machines are another application of refrigeration systems in restaurant equipment. Two kinds of ice machines are flakers (Figure 10-6) and cubers. Most restaurants use ice flakes for salad bar setups, while the ice cubers provide the ice for drinks.

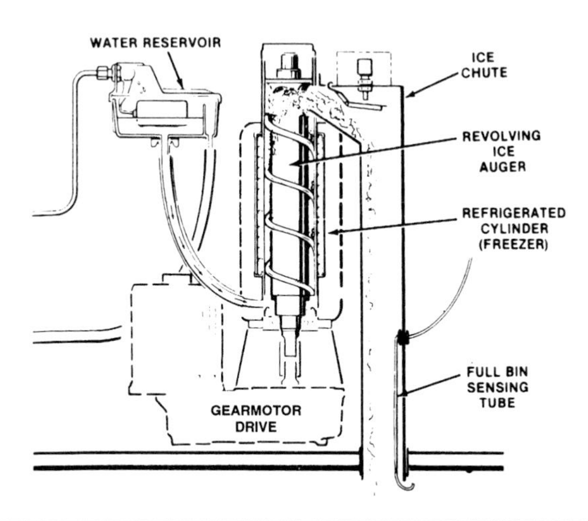

Figure 10-6 An ice flaker uses a refrigerated cylinder in conjunction with an auger. (Courtesy Scotsman Ice Systems)

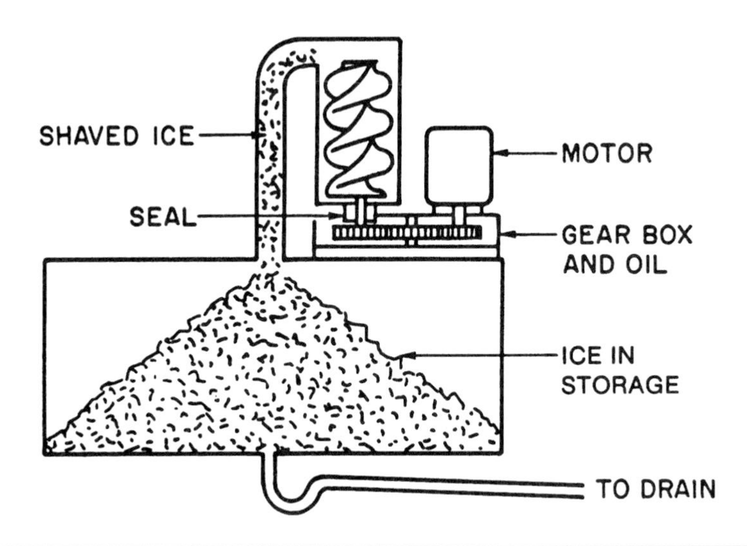

Figure 10-7 After the material is shaved by the auger, it is allowed to drop into the storage bin below.

Flakers – A flaker, such as the one shown in Figure 10-6, uses a refrigerated cylinder in conjunction with an auger. The auger constantly shaves the frozen material from the cylinder and allows it to drop into a bin located underneath the ice machine. (See Figure 10-7.)

Cubers – There are many types of ice cubers on the market. While the basics of refrigeration apply to all different makes and models, the exact method of freezing and harvesting the cubes will differ from one type of machine to another. One system creates ice by flowing a sheet of water over a plate evaporator. It is shown in Figure 10-8.

While you might have been taught that circulating water never freezes, that concept does not apply to an ice machine with a plate-type evaporator. The water does circulate across the surface of the plate, and it does freeze in thin layers.

An interesting fact about commercial ice machines is that the water circulation is what contributes to the clear ice cubes you may notice when you order a drink in a restaurant. While the water is circulating over the plate, many of the minerals that cause cloudy ice cubes, such as the ones produced in a home ice maker, will not stick to the evaporator plate. Instead, these minerals accumulate in the water catch basin of the machine (also referred to as a *sump*). When the unit is cycled into the harvest mode, the sump is flushed automatically. This sends the heavy concentration of minerals down the drain.

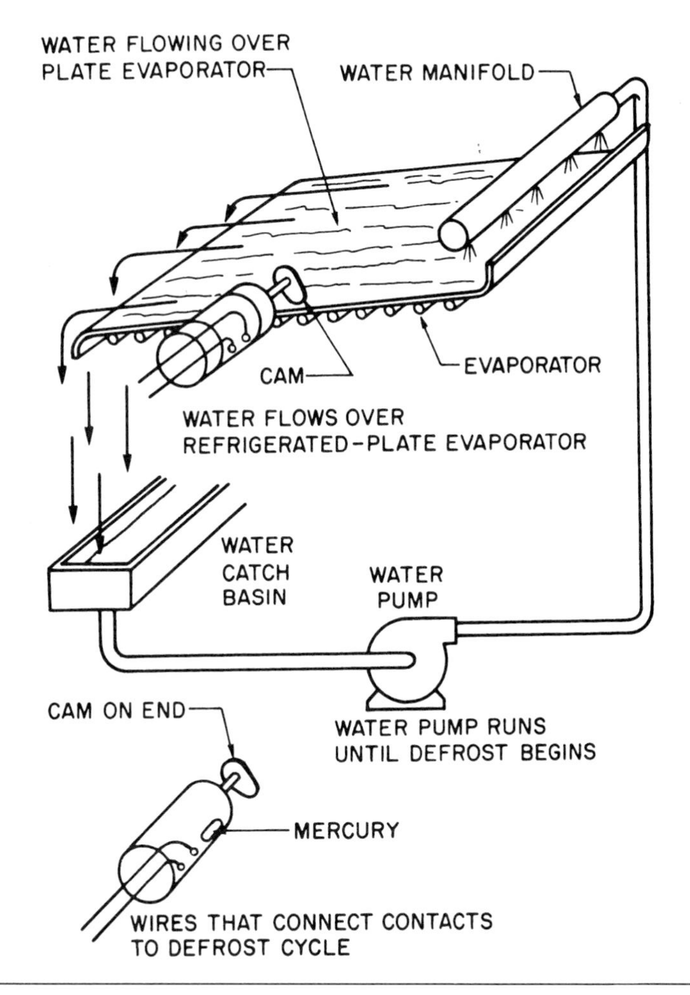

Figure 10-8 Sensors control ice thickness in this machine. The ice thickness switch rotates when ice begins to form on the evaporator plate. The cam touches the ice during rotation (approximately 1 RPM). The switch has mercury contacts in the rear. As the cam touches the ice, it causes the mercury to roll to the back. The contacts are made and defrost begins.

The harvest cycle of an ice cube machine is essentially a hot gas defrost cycle.

In the case of a unit with a plate-type evaporator, when the ice reaches a prescribed thickness, a solenoid redirects the hot gas to the evaporator and the plate warms up. This allows the ice slab to slide off the evaporator and come to rest on a series of wires that act as a cutting grid. A low voltage applied to the cutting grid wires cuts the ice slab into cubes, then drops the cubes into a storage bin below.

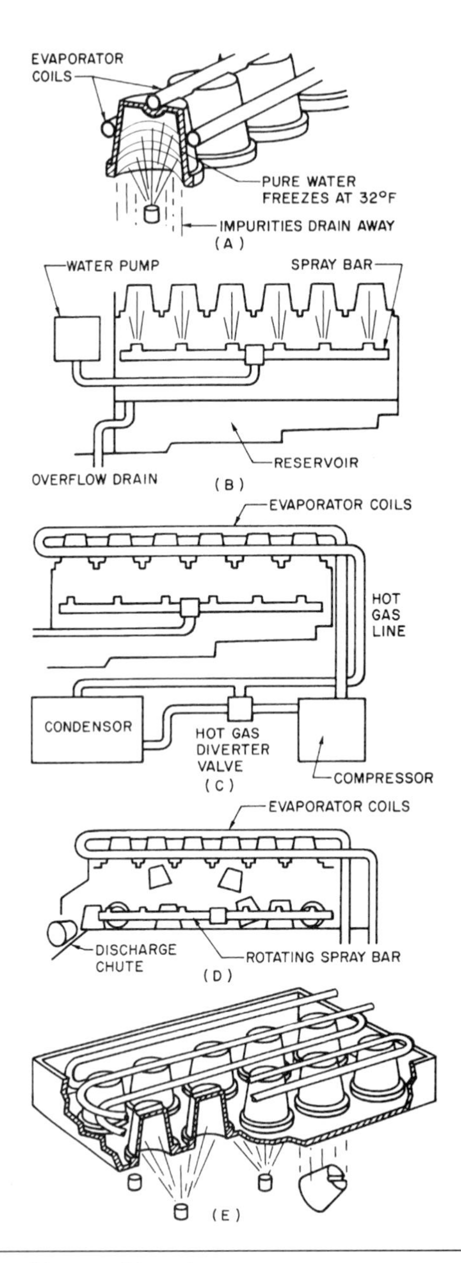

Figure 10-9 One type of ice machine that uses a spray system.

Another type of ice machine is one that uses a spray system, such as the one shown in Figure 10-9.

Still other methods of making ice are the vertical evaporator and the tube system. Both are shown in Figure 10-10.

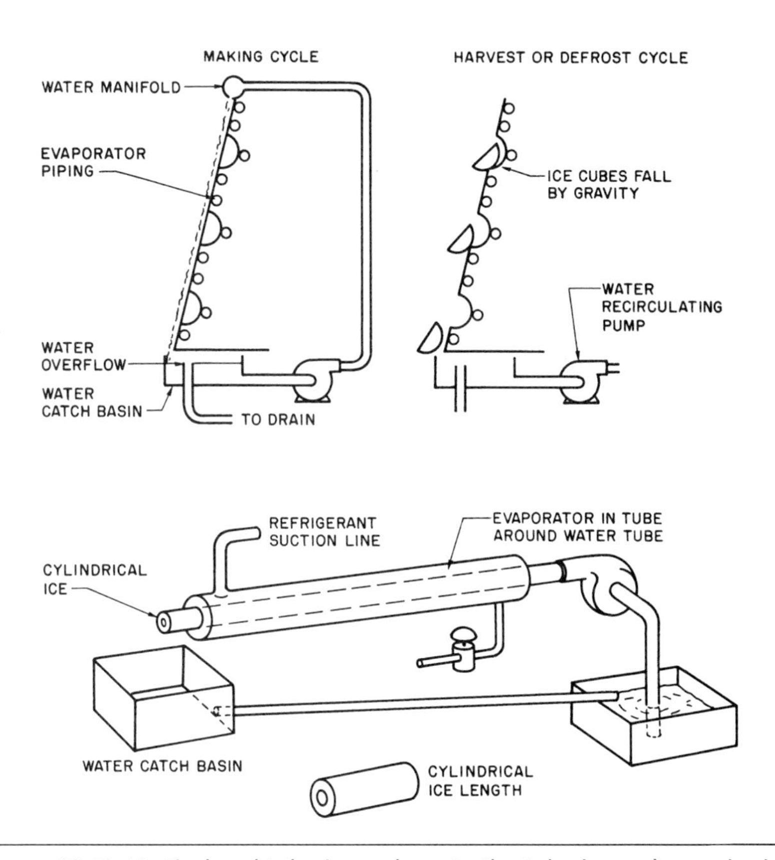

Figure 10-10 Vertical and tube ice makers. In the tube ice maker, as ice is formed on the inside of the tube, the hole in the ice becomes smaller. When a predetermined pump pressure is reached, defrost occurs and the ice shoots out the end, as from a gun. The ice is caught and broken to length, then moved to a bin. In a real machine, the evaporator would normally be wound in a coil.

TROUBLESHOOTING HINT

If you are called in to service an ice machine that does not enter the harvest cycle, the problem could be due to water that is too pure. In many ice machines, manufacturers depend on sensors that use the electrical conductivity of the water to signal the defrost (harvest) mode. If the water is too clean, the harvest cycle will not be initiated. To test for this problem, drop a pinch of salt into the water circulating system. If the harvest cycle initiates, then you know the problem of no harvest is related to water purity. After this problem is diagnosed, you can follow the manufacturer's instructions for solving the situation.

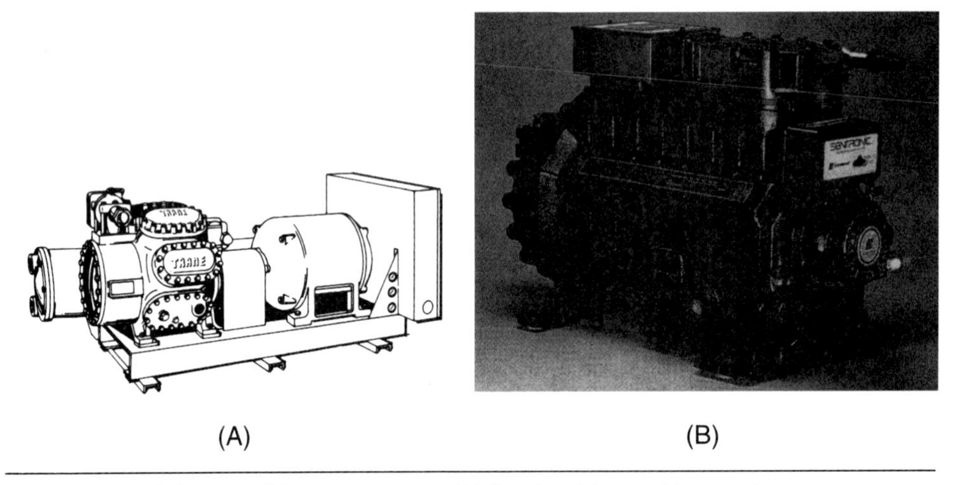

(A) (B)

Figure 10-11 (A) Open-drive compressor. (B) Serviceable semi-hermetic compressor. [(A) Courtesy The Trane Company, La Crosse, WI; (B) courtesy Copeland Corporation]

COMMERCIAL COMFORT COOLING SYSTEMS

One common method of comfort cooling in a commercial application, such as an office complex or apartment building, is the use of what is known as a *hydronic system*. Also known as a *chiller system*, the refrigeration equipment chills water in a central area equipment room, then a series of pipes carry the chilled water to the individual air handlers in each office or apartment. Used in conjunction with what is commonly referred to as a *chiller barrel*, compressors such as the type shown in Figure 10-11 are typical of this application.

REFRIGERATION SYSTEM SERVICE PROCEDURES

Servicing the refrigeration system of commercial equipment differs from servicing domestic or small refrigeration systems in that it will not be necessary to install access valves. Commercial equipment is manufactured with built-in access valves that are either of the Schrader type shown in Figure 10-12, or the serviceable compressor will be equipped with service valves, such as those shown in Figure 10-13.

There are three positions of the compressor service valve: back-seated, mid-seated, and front-seated. Figure 10-13 shows the system access valves in the mid-seated position. In this position, there is a path for the gauges to connect to the refrigeration system components. The mid-seated position is used when evacuating, recharging, or diagnosing a system.

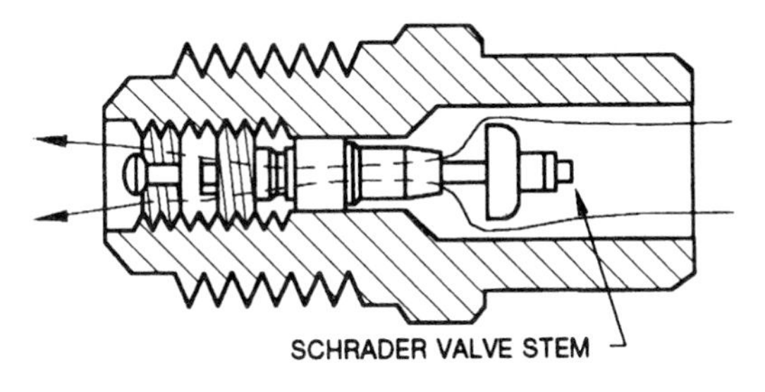

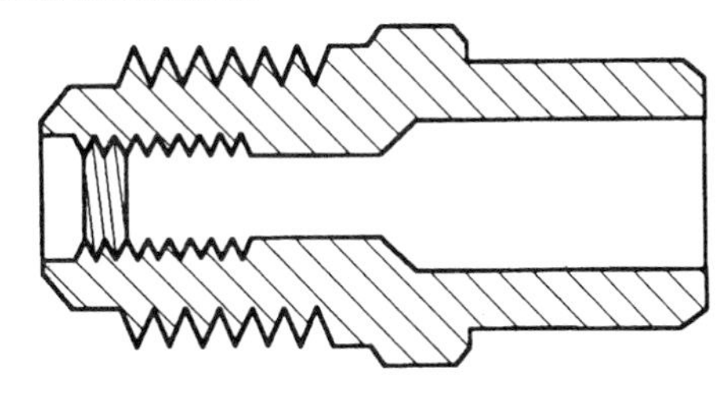

Figure 10-12 The Schrader valve stem may be removed in some equipment to improve the gas flow.

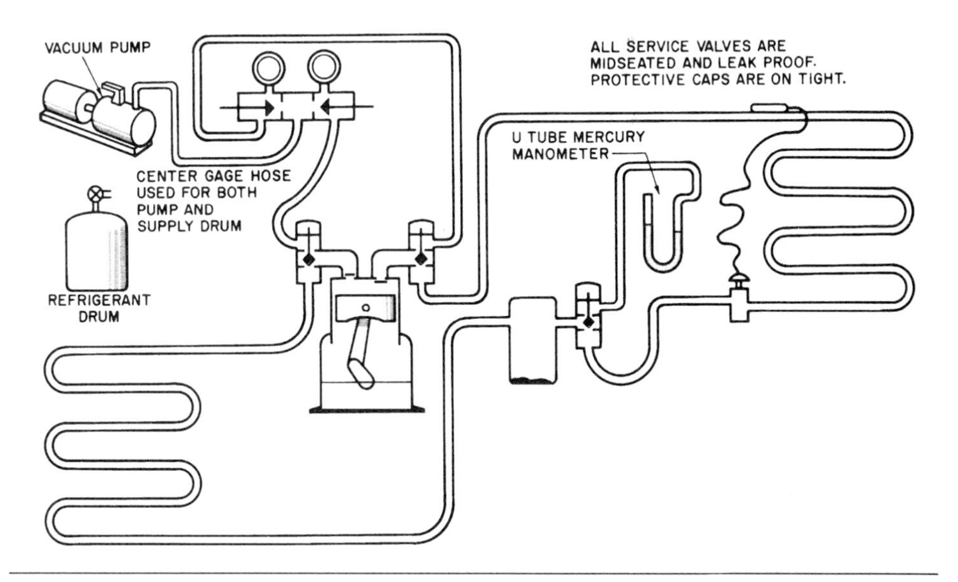

Figure 10-13 A serviceable compressor comes equipped with service valves.

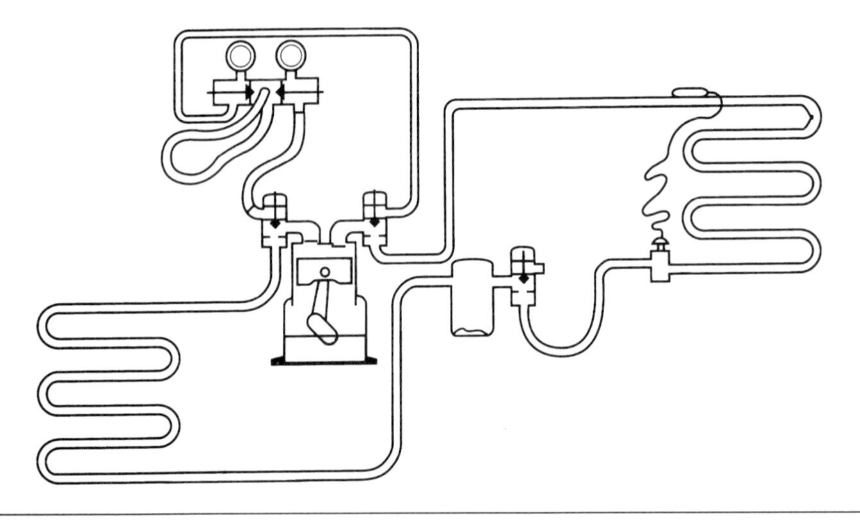

Figure 10-14 When the compressor service valves are in their back-seated position, the system can operate normally, but the gauges cannot read the system pressures.

When the access valves are in their back-seated position, as shown in Figure 10-14, the system can operate normally but the gauges cannot read the system pressures.

When the access valves are in their front-seated position, such as shown in Figure 10-15, the refrigeration system will not operate because the compressor is closed off, and the condenser and evaporator are isolated.

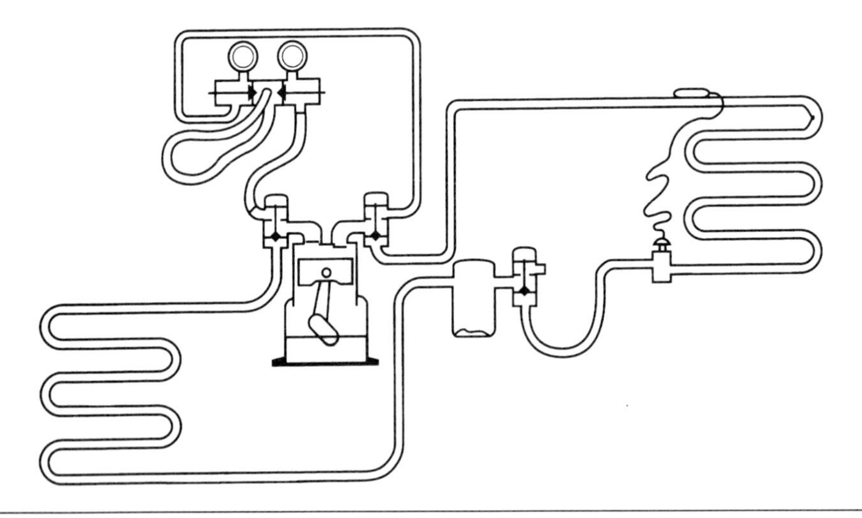

Figure 10-15 The refrigeration system cannot operate with the compressor service valves in their front-seated position.

In order to diagnose an operating system, the compressor access valves must be in the mid-seated position.

When accessing a sealed system for diagnosis and troubleshooting, you need a working knowledge of gauges, an understanding of the temperature-pressure relationship in refrigerants, and an understanding of temperature/pressure (T/P) charts.

The idea of a temperature-pressure relationship in a refrigerant is easy to understand: if the temperature goes up, the pressure of the refrigerant goes up, and if the temperature goes down, the pressure of the refrigerant goes down. It does not matter if the drum contains one ounce or one hundred pounds of refrigerant—the temperature-pressure relationship still applies. This concept is illustrated in Figure 10-16.

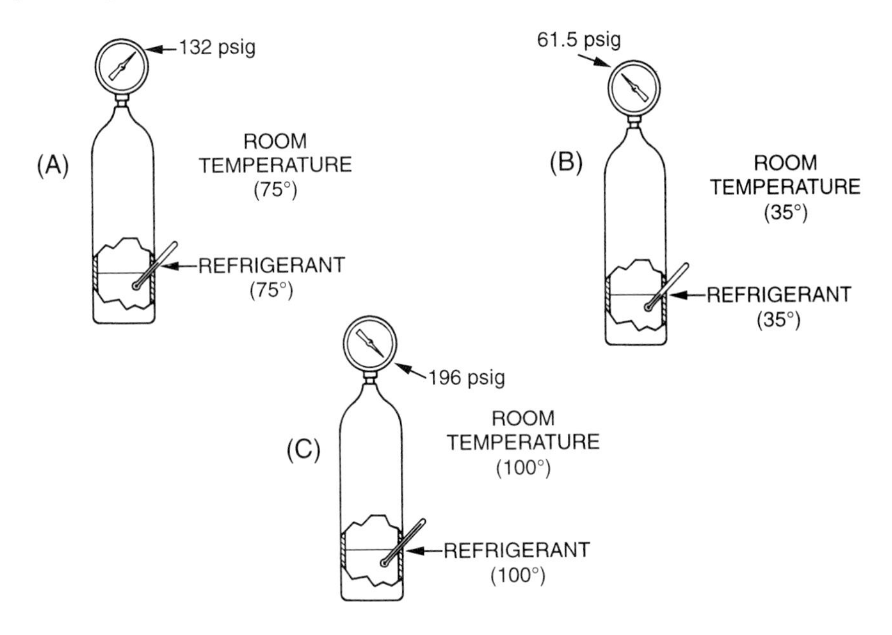

Figure 10-16 (A) The drum of HCFC-22 is left in a 75°F room until the drum and its contents are at room temperature. The drum contains a partial liquid, partial vapor mixture, which, when both become room temperature, will be said to be in *equilibrium*; no more temperature changes will be taking place. At this time, the drum pressure, 132 psig, will correspond to the drum temperature of 75°F. (B) The drum is moved into a walk-in cooler and left until the drum and its contents become the same as the inside of the cooler at 35°F. Until the drum and its contents get to 35°F, some of the vapor in the drum will be changing to a liquid and reducing the pressure. Soon the drum pressure will correspond to the room temperature of 35°F, 61.5 psig. (C) The drum is moved into a 100°F room and allowed to reach the point of equilibrium at 100°F, 196 psig. The pressure rise is due to some of the liquid refrigerant boiling into a vapor and increasing the total drum pressure.

For example, a drum of HCFC-22 (quantity of material notwithstanding) can be placed in an ambient temperature of 35°F, 75°F, and 100°F, and the pressure in the drum will change accordingly.

This concept can be further illustrated with a temperature/pressure chart such as the one shown in Figure 10-17.

With this chart, you can understand how the pressure reading of 61.5 psig was accomplished in Figure 10-16.

STEP ONE: Working with the T/P chart, find 35°F in the appropriate temperature column.

STEP TWO: Identify the column of pressure readings under refrigerant 22.

STEP THREE: Reading across from 35°F to the refrigerant 22 column will give you a reading of 61.5 psig.

The T/P chart is also used to evaluate an operating refrigeration system, and provides a basis to work from when diagnosing and troubleshooting. To decide whether a refrigeration system is performing properly, you must first know the normal operating conditions.

TEMPERATURE	REFRIGERANT				TEMPERATURE	REFRIGERANT				TEMPERATURE	REFRIGERANT			
°F	12	22	134a	502	°F	12	22	134a	502	°F	12	22	134a	502
−60	19.0	12.0		7.2	12	**15.8**	**34.7**	**13.2**	**43.2**	42	**38.8**	**71.4**	**37.0**	**83.8**
−55	17.3	9.2		3.8	13	**16.4**	**35.7**	**13.8**	**44.3**	43	**39.8**	**73.0**	**38.0**	**85.4**
−50	15.4	6.2		0.2	14	**17.1**	**36.7**	**14.4**	**45.4**	44	**40.7**	**74.5**	**39.0**	**87.0**
−45	13.3	2.7		**1.9**	15	**17.7**	**37.7**	**15.1**	**46.5**	45	**41.7**	**76.0**	**40.1**	**88.7**
−40	11.0	**0.5**	14.7	**4.1**	16	**18.4**	**38.7**	**15.7**	**47.7**	46	**42.6**	**77.6**	**41.1**	**90.4**
−35	8.4	**2.6**	12.4	**6.5**	17	**19.0**	**39.8**	**16.4**	**48.8**	47	**43.6**	**79.2**	**42.2**	**92.1**
−30	5.5	**4.9**	9.7	**9.2**	18	**19.7**	**40.8**	**17.1**	**50.0**	48	**44.6**	**80.8**	**43.3**	**93.9**
−25	2.3	**7.4**	6.8	**12.1**	19	**20.4**	**41.9**	**17.7**	**51.2**	49	**45.7**	**82.4**	**44.4**	**95.6**
−20	**0.6**	**10.1**	3.6	**15.3**	20	**21.0**	**43.0**	**18.4**	**52.4**	50	**46.7**	**84.0**	**45.5**	**97.4**
−18	**1.3**	**11.3**	2.2	**16.7**	21	**21.7**	**44.1**	**19.2**	**53.7**	55	**52.0**	**92.6**	**51.3**	**106.6**
−16	**2.0**	**12.5**	0.7	**18.1**	22	**22.4**	**45.3**	**19.9**	**54.9**	60	**57.7**	**101.6**	**57.3**	**116.4**
−14	**2.8**	**13.8**	**0.3**	**19.5**	23	**23.2**	**46.4**	**20.6**	**56.2**	65	**63.8**	**111.2**	**64.1**	**126.7**
−12	**3.6**	**15.1**	**1.2**	**21.0**	24	**23.9**	**47.6**	**21.4**	**57.5**	70	**70.2**	**121.4**	**71.2**	**137.6**
−10	**4.5**	**16.5**	**2.0**	**22.6**	25	**24.6**	**48.8**	**22.0**	**58.8**	75	**77.0**	**132.2**	**78.7**	**149.1**
−8	**5.4**	**17.9**	**2.8**	**24.2**	26	**25.4**	**49.9**	**22.9**	**60.1**	80	**84.2**	**143.6**	**86.8**	**161.2**
−6	**6.3**	**19.3**	**3.7**	**25.8**	27	**26.1**	**51.2**	**23.7**	**61.5**	85	**91.8**	**155.7**	**95.3**	**174.0**
−4	**7.2**	**20.8**	**4.6**	**27.5**	28	**26.9**	**52.4**	**24.5**	**62.8**	90	**99.8**	**168.4**	**104.4**	**187.4**
−2	**8.2**	**22.4**	**5.5**	**29.3**	29	**27.7**	**53.6**	**25.3**	**64.2**	95	**108.2**	**181.8**	**114.0**	**201.4**
0	**9.2**	**24.0**	**6.5**	**31.1**	30	**28.4**	**54.9**	**26.1**	**65.6**	100	**117.2**	**195.9**	**124.2**	**216.2**
1	**9.7**	**24.8**	**7.0**	**32.0**	31	**29.2**	**56.2**	**26.9**	**67.0**	105	**126.6**	**210.8**	**135.0**	**231.7**
2	**10.2**	**25.6**	**7.5**	**32.9**	32	**30.1**	**57.5**	**27.8**	**68.4**	110	**136.4**	**226.4**	**146.4**	**247.9**
3	**10.7**	**26.4**	**8.0**	**33.9**	33	**30.9**	**58.8**	**28.7**	**69.9**	115	**146.8**	**242.7**	**158.5**	**264.9**
4	**11.2**	**27.3**	**8.6**	**34.9**	34	**31.7**	**60.1**	**29.5**	**71.3**	120	**157.6**	**259.9**	**171.2**	**282.7**
5	**11.8**	**28.2**	**9.1**	**35.8**	35	**32.6**	**61.5**	**30.4**	**72.8**	125	**169.1**	**277.9**	**184.6**	**301.4**
6	**12.3**	**29.1**	**9.7**	**36.8**	36	**33.4**	**62.8**	**31.3**	**74.3**	130	**181.0**	**296.8**	**198.7**	**320.8**
7	**12.9**	**30.0**	**10.2**	**37.9**	37	**34.3**	**64.2**	**32.2**	**75.8**	135	**193.5**	**316.6**	**213.5**	**341.2**
8	**13.5**	**30.9**	**10.8**	**38.9**	38	**35.2**	**65.6**	**33.2**	**77.4**	140	**206.6**	**337.2**	**229.1**	**362.6**
9	**14.0**	**31.8**	**11.4**	**39.9**	39	**36.1**	**67.1**	**34.1**	**79.0**	145	**220.3**	**358.9**	**245.5**	**385.0**
10	**14.6**	**32.8**	**11.9**	**41.0**	40	**37.0**	**68.5**	**35.1**	**80.5**	150	**234.6**	**381.5**	**262.7**	**408.4**
11	**15.2**	**33.7**	**12.5**	**42.1**	41	**37.9**	**70.0**	**36.0**	**82.1**	155	**249.5**	**405.1**	**280.7**	**432.9**

Note: Bold figures in the refrigerant columns represent gauge pressure; remining figures in those columns represent vacuum pressure.

Figure 10-17 Pressure and temperature relationship chart in in. Hg vacuum or psig.

For example, suppose that you connect your gauges to a refrigeration system that is operating in a 75°F ambient temperature, and the refrigerant being used is HCFC-22. When you connect your high side hose, the gauge indicates a pressure reading of 210 psig. Now the question is, how do you decide whether this reading is normal or abnormal? The answer can be found in the T/P chart.

To use the T/P chart, follow these steps:

STEP ONE: Record the ambient temperature and find it in the appropriate temperature column. (In this case, 75°F).

STEP TWO: Add 30°F to the ambient reading and find it in the appropriate column (75°F + 30°F = 105°F). This is the rule-of-thumb number that is added to calculate the correct high side operating pressure of a refrigeration system with an air-cooled condenser. The reason the addition must be made is due to the heat generated in the refrigeration system in the process of operating the compressor. Friction (and heat) from bearings and the heat of compression are the factors that make this addition necessary.

STEP THREE: Read across from the final calculated temperature to the appropriate refrigerant column (in our example, refrigerant 22), and the chart will offer the figure that represents what the high side operating pressure of a system should be. Since the result in this case is 210.8 psig, we know that the refrigeration system in question is operating properly. The reading found when the high side hose was connected matches the reading offered by the chart.

Gauges used to access and read pressures on refrigeration systems can vary from the standard three-hose compound gauge system shown in Figure 10-18. Also available are the four-hose system shown in Figure 10-19, and the digital gauges shown in Figure 10-20.

In addition to reading pressures in an operating system, a refrigerant gauge can also be a further illustration of the pressure/temperature relationship. Along with the pressure scale printed on the face of the gauge, several scales are also shown, and these scales represent the temperature of a refrigerant in a system. In the gauge shown in Figure 10-21, the three refrigerants represented on the additional scales are R-12 (CFC-12), R-22 (HCFC-22) and R-502 (CFC-502), all refrigerants that can be found in commercial applications.

Figure 10-18 Two-valve gauge manifold. (By Bill Johnson)

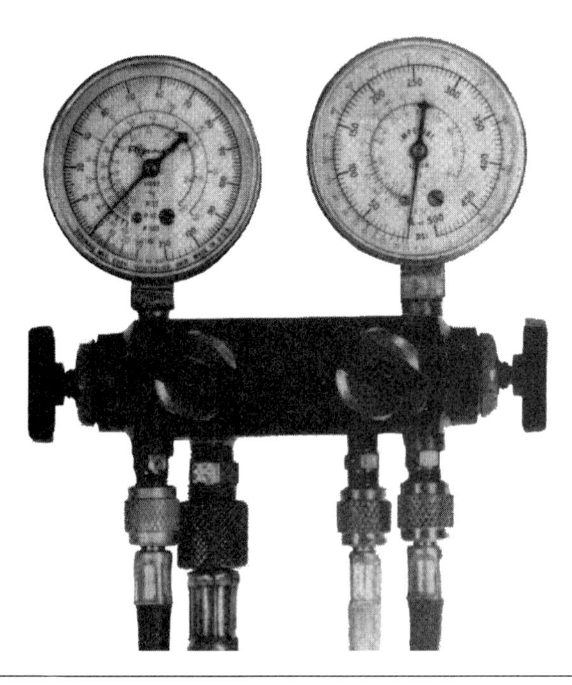

Figure 10-19 Four-valve gauge manifold. (By Bill Johnson)

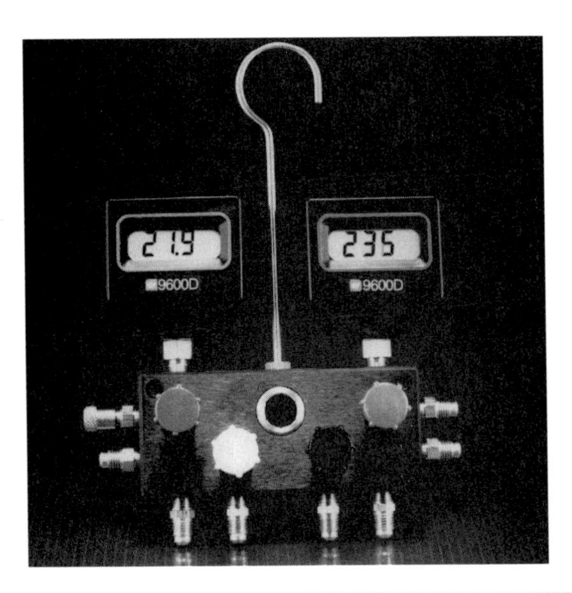

Figure 10-20 A digital gauge set. (Courtesy TIF Instruments, Inc.)

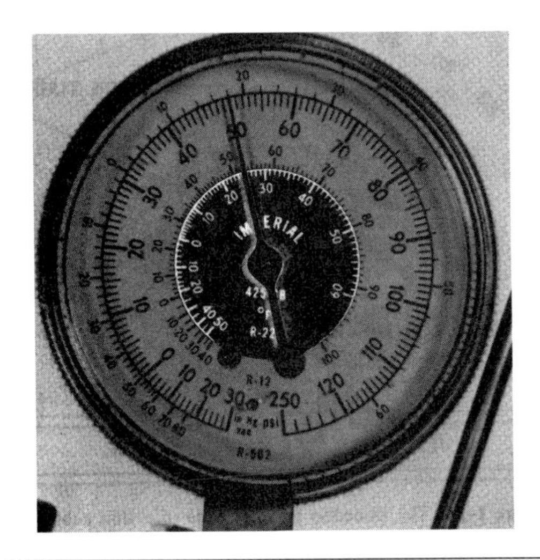

Figure 10-21 Along with the pressure scale printed on the face of the gauge, scales that represent the temperature of a refrigerant in a system are also shown. (By Bill Johnson)

As an example, the R-22 (HCFC-22) scale (which is the innermost scale of numbers on the face of the gauge) is highlighted in Figure 10-21 and the needle is indicating a reading of approximately 25 on that scale. This represents a temperature of 25°F.

The needle is also indicating a reading of approximately 50 on the pressure scale, representing a pressure of approximately 50 psig.

What this means is that if this particular gauge was connected to the low side of a normally operating R-22 (HCFC-22) system, the reading of approximately 50 psig would indicate a refrigerant temperature of approximately 25°F, or the temperature of 25°F would translate into a pressure of 50 psig.

You can confirm these readings by referring back to the temperature/pressure chart in Figure 10-17. Finding 25°F in the appropriate column and reading across to the refrigerant 22 scale, you will find that it corresponds to a pressure of 48.8 psig.

Conversely, you could find a reading of 48.8 psig (approximately 50 psig), and by reading across to the temperature scale, you would find that it corresponds to a temperature of 25°F.

ELECTRICAL SYSTEM SERVICE PROCEDURES

The use of schematic and pictorial diagrams is an important part of the service and troubleshooting process for commercial equipment. Schematic diagrams show each circuit in an organized line-by-line layout, and pictorial diagrams show the physical location of the components in the system.

Start Devices

Unlike small refrigeration systems, a popular start device in commercial equipment is the potential relay (refer to Unit 6). It is used in conjunction with a start capacitor. Contactors and "mag" starters are also used as motor operating devices in commercial equipment. In many cases, the control voltage used to operate the contactor or starter is the same voltage applied to the operating components in the system.

Pressure Switches

Some commercial equipment, such as walk-in coolers, may be controlled by a pressure switch on the low side of the system, rather than a cold control such as those found on smaller refrigeration systems.

The pressure switch control system works on the principle of the temperature/pressure relationship within the refrigerant. With the sensing portion of the control connected to the low pressure side of the system, it senses a drop in pressure as the temperature of the refrigerant drops during a run cycle. When the design temperature is reached, the switching portion of the pressure switch opens, breaking the circuit to the compressor.

As the temperature inside the cabinet rises, the refrigerant pressure also rises, causing the switch to close. This completes the circuit to the compressor for another run cycle.

Oil pressure safety switches and high pressure switches used as protective devices are also common in commercial refrigeration equipment.

THE AIR HANDLING SYSTEM

Due to its heavier system load, commercial refrigeration equipment requires either a forced-air system for both the evaporator and condenser section of the equipment, or water is used as a heat transfer agent to cool the condenser. Whatever the design, a forced-air system rather than a static system is used.

Any restriction of air flow over either the evaporator or condenser section of the system will have a dramatic effect on efficiency.

For example, in a system that uses multiple evaporator fans, the failure of one fan will cause insufficient air flow that will ultimately lead to insufficient heat transfer and higher operating temperatures. A lack of proper air flow over the evaporator can also cause liquid slugging in the compressor. This is due to the fact that the proper level of heat absorption does not take place in the evaporator, allowing some of the refrigerant to pass through without changing state.

Improper air flow through an evaporator can also be caused by a defrost problem or a lack of proper maintenance. A dirty coil or a failure of the defrost system will restrict the air flow over the evaporator coil.

In the event of defrost system failure, the problem must be solved and the evaporator must be cleared of frost. One way to accomplish this is to use a heat gun such as the one shown in Figure 10-22.

Figure 10-22 A heat gun used for defrosting frost-clogged evaporator coils. (Courtesy Robinair Division, SPX Corporation)

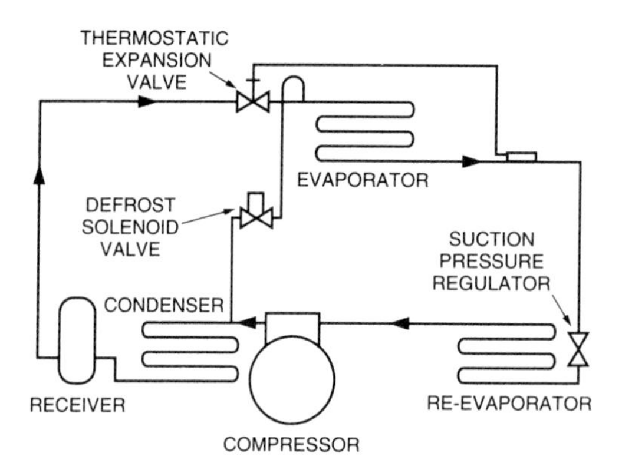

Figure 10-23 An illustration of the physical layout of a hot gas defrost system. When energized, the solenoid allows hot gas from the compressor to flow directly to the evaporator. During a refrigeration cycle, the solenoid is closed, and refrigerant flow follows the normal path through the refrigeration system components.

PROBLEM 4 *A WALK-IN FREEZER WITH A HOT GAS DEFROST SYSTEM IS FROSTING UP AND THE TEMPERATURE IS RISING*

Upon your arrival, you find the following conditions:

1. The compressor, condenser fan, and evaporator fan are all running normally.
2. The evaporator is clogged with frost.

This system (shown in Figure 10-23) uses a defrost solenoid valve. It is a normally closed (N.C.) valve, meaning that unless power is applied to its coil, the valve is in the closed position.

The hot gas defrost system is shown in Figure 10-24. All components in the system operate on line voltage, and the hot gas solenoid valve is energized when the timer advances to the defrost mode.

The steps to troubleshoot this problem are as follows:

STEP ONE: Locate the timer assembly and manually rotate the mechanism to the defrost mode. (You know it is not in the defrost mode because the evaporator fan motor is running. The N.C. contacts in the timer are allowing a circuit to the evaporator fan motor. In the defrost mode, these contacts would open as the normally open or N.O. contacts to the hot gas solenoid valve close.)

When you reach the defrost mode of the timer, the evaporator fan motor stops. Proceed to step two.

STEP TWO: With the timer in the defrost mode, test for voltage at the wire connections of the hot gas solenoid valve. This test shows that the equipment's applied line voltage is present at the solenoid valve electrical connections.

Which component must be replaced to solve the problem?

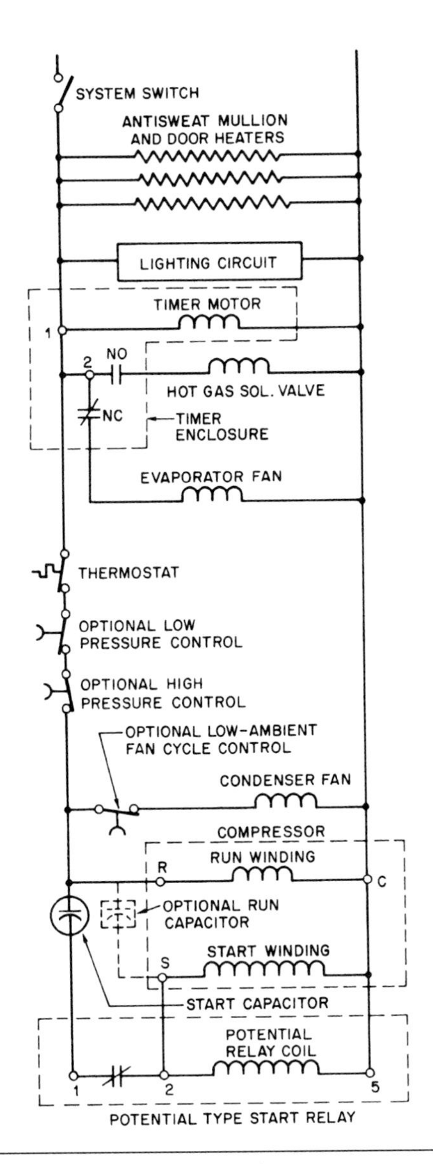

Figure 10-24 Wiring diagram of a hot gas defrost system.

PROBLEM 5 *A REACH-IN FREEZER IS FROSTING UP AND THE TEMPERATURE IS RISING*

In this situation, defrost is accomplished with electric heaters mounted to the evaporator, rather than with a hot gas solenoid valve. The schematic diagram in Figure 10-25 shows this electrical system.

Upon your arrival, you find the following conditions:

1. The compressor, condenser fan motor, and evaporator fan motor are all running normally.
2. The evaporator is clogged with frost.

The steps to troubleshoot this problem are as follows:

STEP ONE: Locate the timer assembly and manually advance it to the defrost mode. When you advance the timer, the evaporator fan motor, compressor, and condenser fan motor stop. Proceed to step two.

STEP TWO: Using a voltmeter, test for the applied equipment voltage at the electrical connections of the defrost heater. No voltage is read at this point. Proceed to step three.

STEP THREE: With the power disconnected, use an ohmmeter to test for continuity across the limit switch electrical connections. The body of the limit switch is buried in the frost. The ohmmeter shows no continuity across the limit switch.

Which component must be replaced to accomplish the repair on this unit?

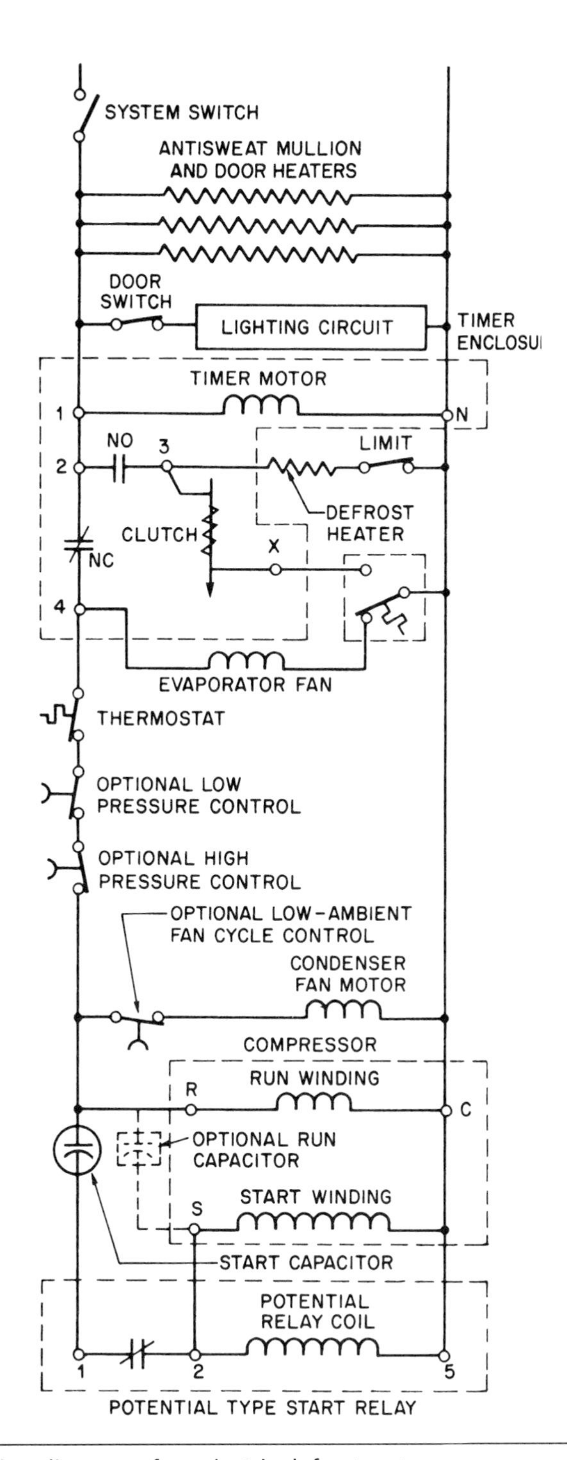

Figure 10-25 Wiring diagram of an electric defrost system.

PROBLEM 6 *A DAIRY DISPLAY CASE IS NOT COOLING PROPERLY*

Upon your arrival to service this self-contained dairy display case, you find the following conditions:

1. The compressor, condenser fan, and evaporator fans are running.
2. The temperature in the unit is 10°F higher than the desired temperature.
3. There is a partial frost pattern on the evaporator.

The steps to troubleshoot this problem are as follows:

STEP ONE: Determine the type of refrigerant used in the equipment, and the normal operating temperature of the evaporator coil. (In this case, R-12 (CFC-12) is the refrigerant and the coil temperature is supposed to be 20°F.) Proceed to step two.

STEP TWO: Connect your gauges to the high side and low side access valves. Proceed to step three.

STEP THREE: Determine the ambient temperature. The temperature reading is 75°F. Proceed to step four.

STEP FOUR: Read the high and low side gauges. The readings are as follows: high side, 85 psig; low side, 15 psig.

With these factors determined, what is the cause of the problem and what are the steps necessary to complete the repair?

UNIT TEN SUMMARY

Commercial refrigeration systems include those used in retail establishments, such as grocery stores, convenience stores, and restaurants, as well as those used in large comfort cooling systems, such as office buildings and apartment complexes.

Grocery stores make extensive use of refrigeration equipment for food preservation and merchandising. The display cases are located throughout the store, and the compressors are located in a central equipment room. Some grocery stores have a heat reclaim system that makes use of the heat during the winter months, using an air handling system to circulate warm air through the store.

Closed display cases, open display cases, and walk-in coolers and freezers are used in grocery stores.

In restaurants, reach-in refrigerators, reach-in freezers, and walk-in coolers and freezers are used for food preservation. Commercial ice machines, some that make cubes and others that make flaked ice, are also used extensively in the restaurant industry.

Hydronic systems, also known as *chiller systems,* are used in apartment complexes and office buildings. A central equipment room houses the refrigeration system and chilled water is routed to individual coils in the apartments or offices.

Unlike small refrigeration systems, commercial equipment comes equipped with factory-installed service valves. Either the system will be designed to use a serviceable compressor that is equipped with service valves, or Schrader-type service valves will be installed at the factory.

The three positions of the compressor service valve are back-seated, front-seated, and mid-seated. In the mid-seated position, you can connect gauges and evaluate system operation. In the back-seated position, the system is operational, and in the front-seated position, the compressor is isolated from the rest of the refrigeration system.

When working with commercial refrigeration systems, the T/P chart is often used to determine whether or not a system is functioning normally.

Schematic and pictorial diagrams are an important part of the service procedures regarding commercial refrigeration equipment. Schematic diagrams show the circuit, and pictorial diagrams show the physical layout of the components.

The potential relay is a popular start device in commercial equipment. In many cases, commercial refrigeration systems are controlled and protected by pressure switches.

Comfort Cooling and Heating Systems

PACKAGED UNITS

One of the most common types of combination heating/cooling units found in residential applications is the packaged unit. This type of unit may use electric heat, gas heat, or it may be a heat pump system in which heating is accomplished through what is known as *reverse cycle refrigeration*.

Whatever the process for accomplishing the heating cycle, the cooling mode is accomplished through the use of refrigeration system components applied to a comfort cooling system.

The compressor, condenser (often referred to as the *outdoor coil*), evaporator *(indoor coil)*, and metering device, which can range from a multiple capillary tube system to a thermostatic expansion valve, make up the refrigeration system.

The air in the conditioned space is circulated through the indoor coil, and the refrigerant circulating through the coil absorbs heat from the air.

In the case of a cooling/gas heat unit, such as the one shown in Figure 11-1, the system contains a heating section made up of gas burners, a gas valve, and a heat exchanger section in addition to the refrigeration system.

The flow of air to and from the conditioned space is always through the heat exchanger, even in the cooling mode. In the heating mode, the air flow path is through the indoor coil. However, in the heating mode, the refrigeration system is not operating, and in the cooling mode, the heating system is not in operation.

Figure 11-1 A packaged A/C unit. (Courtesy Rheem Air-Conditioning Division)

TROUBLESHOOTING HINT

When responding to a service call on a packaged unit in which the complaint is diminished operating capacity, check the condition of the indoor coil for dirt buildup and eliminate that as a possibility before pursuing a possible problem in the refrigeration system. Checking the contact side of the indoor coil may require that you take the top off the unit or even disconnect it from the ductwork and move it slightly in order to visually inspect the coil, but it is worth the trouble. A lot of unnecessary work can be avoided by first inspecting the indoor coil.

In the case of a unit that uses a refrigeration system in conjunction with electric heat, resistance heaters are located in the path of air flow to the conditioned space. A packaged unit that uses electric heat is similar in appearance to the gas unit as far as cabinet configuration and duct connections, but one thing that you will immediately recognize is the absence of a vent system (see Figure 11-2).

A packaged unit that uses a gas-fired heating system will either have a power vent system of some kind with the vent outlet mounted to the side of the cabinet assembly (shown in Figure 11-1) or a vent tower of some type will be located on the top of the cabinet, such as shown in Figure 11-3.

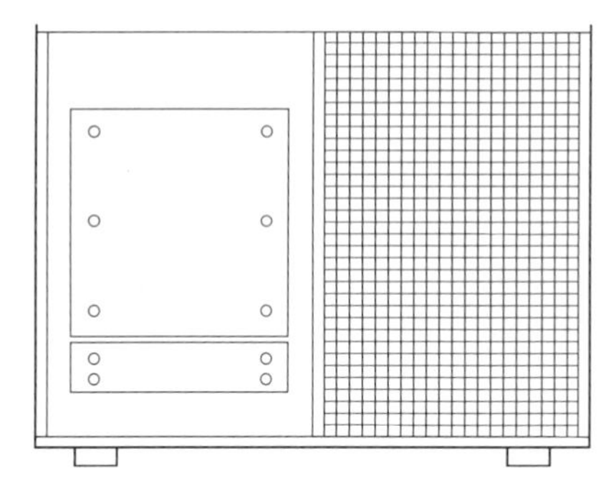

Figure 11-2 In a refrigeration system with electric heat, a vent system is not necessary.

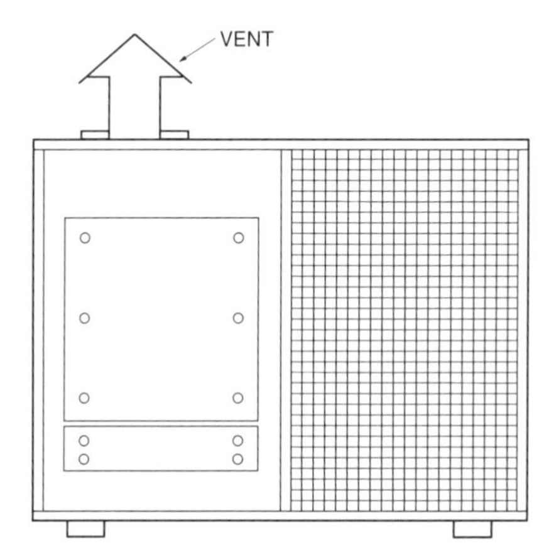

Figure 11-3 Many gas packs have a vent tower located on the top of the cabinet.

The packaged heat pump is similar in appearance to the electric heat packaged unit since heat pumps also do not require vents. Figure 11-4 shows a packaged heat pump unit.

Another feature of heat pumps is the supplementary heating system. In addition to the reverse cycle heating system, supplementary heat is provided by resistance-type heating elements of the same type found in electric furnaces and electric heat packaged units. A quick check to determine the difference between a heat pump and an electric heat unit is the presence of a reversing valve and accumulator located near the refrigeration system compressor.

The reversing valve in a heat pump is the component that redirects the refrigerant flow when the unit controls switch the cycle from heating to cooling or vice versa. The two different refrigerant flow patterns in a heat pump are shown in Figure 11-5.

Figure 11-4 Packaged unit heat pump. (Courtesy Rheem Air-Conditioning Division)

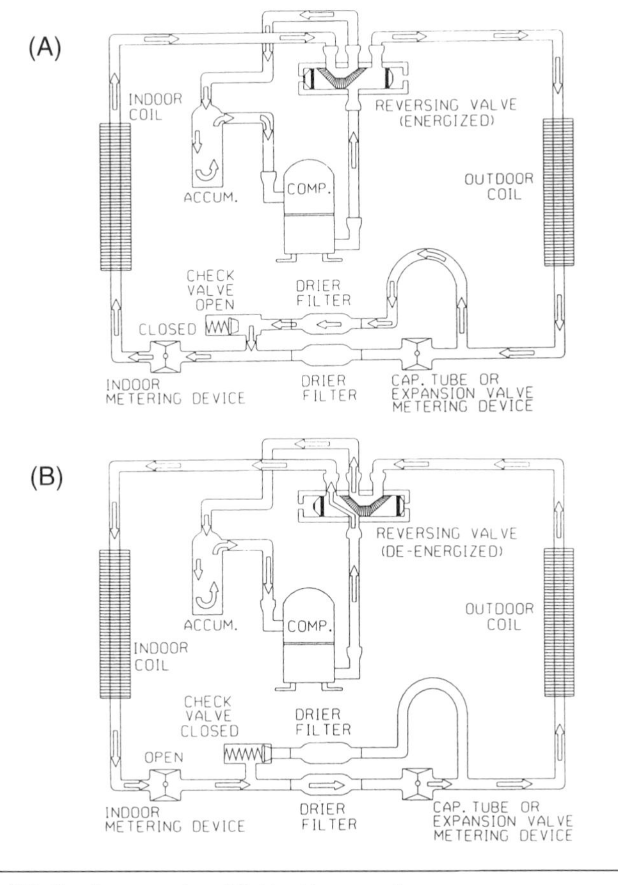

Figure 11-5 (A) Cooling mode. (B) Heating mode.

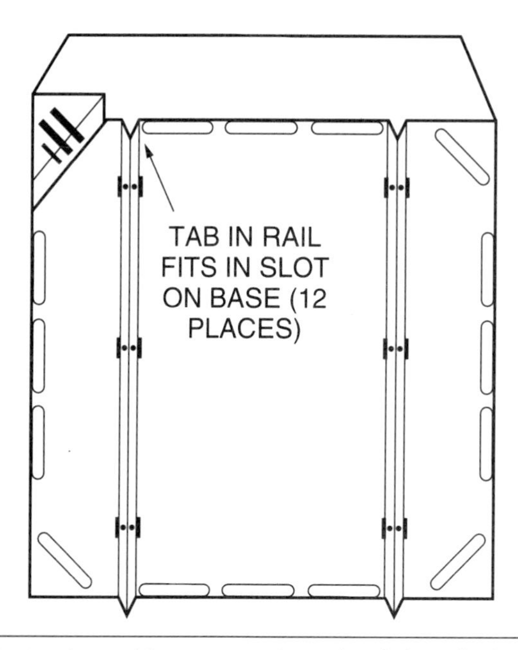

Figure 11-6 A curb stop is used in many packaged unit installations.

Packaged air conditioning systems can be set on a pad at ground level next to a building, in which case a ductwork system will handle the air flow to and from the conditioned space. Roof mounting of packaged units is also common. In many cases, a roof mounting system known as a *curb stop* is used (see Figure 11-6). This allows the ductwork connections to be made on the underside of the unit. Roof-mounted units can also be connected as shown in Figure 11-7. In this installation, two sheet metal elbows are used to provide the supply and return air flow to and from the conditioned space.

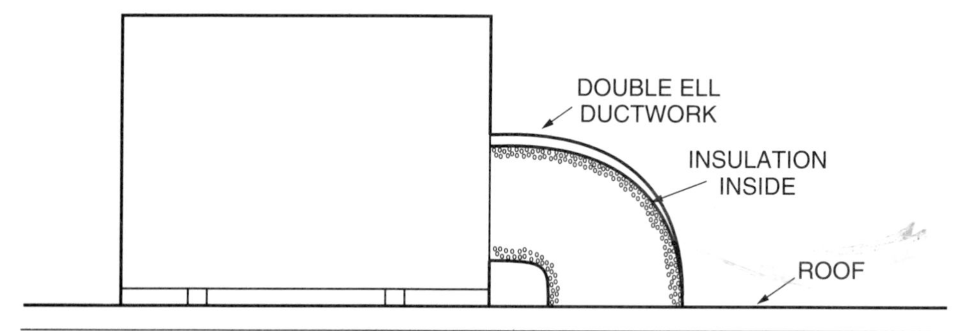

Figure 11-7 Sheet metal ductwork in the form of a 90° elbow carries the air to and from the conditioned space below. The duct is insulated on the inside.

TROUBLESHOOTING HINT

If you suspect an air flow problem in a roof-mounted unit with double ell ductwork, check for loose insulation on the inside of the ductwork. Hanging insulation could cause a restriction in the air flow.

SPLIT SYSTEMS

A combination heating/cooling system that uses an indoor furnace (gas or electric) and an outdoor condensing unit is known as a *split system*. The condensing unit contains the compressor and condenser, while the evaporator section of the cooling system is located in the ductwork inside. A typical split system is shown in Figure 11-8.

The indoor coil of a gas or electric heat split system is often constructed in the shape of an "A" and is referred to as an *A-coil*. Split systems used in heat pump applications generally use a flat indoor coil mounted in a slant position in the ductwork.

GAS FURNACES

Gas furnaces are designed in one of two air flow configurations: downflow or upflow.

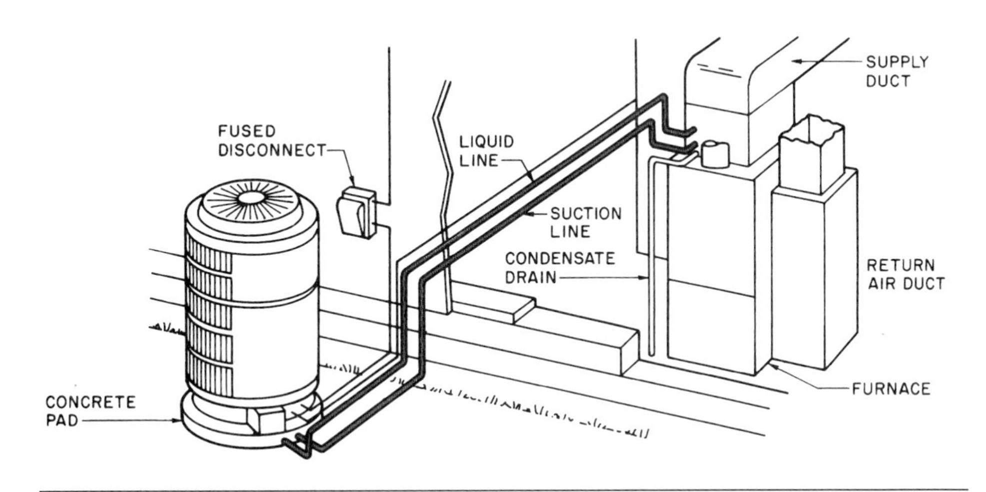

Figure 11-8 Complete installation with the piping shown. The evaporator is piped to the condensing unit on the outside of the structure.

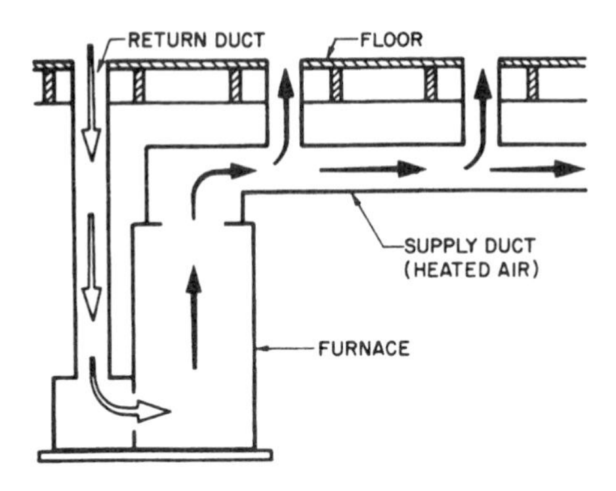

Figure 11-9 The air flow pattern of an upflow furnace in a basement.

Upflow Furnaces

The upflow furnace is the most popular design. Figure 11-9 shows the air flow pattern of an upflow furnace installed in a basement location.

Not all upflow furnaces are installed in basements. In ranch style homes without basements, the upflow furnace will be found in a hallway closet, and the supply ducts are routed in a crawl space above the ceiling. Whatever the application, the return air flow to an upflow gas furnace is through the bottom or side of the furnace cabinet, and the warm air discharge is through the top.

Downflow Furnaces

A downflow gas furnace differs from the upflow type in that the pattern of warm air discharge is out the bottom of the cabinet and the return is through the top. Figure 11-10 shows the air flow pattern of a downflow furnace.

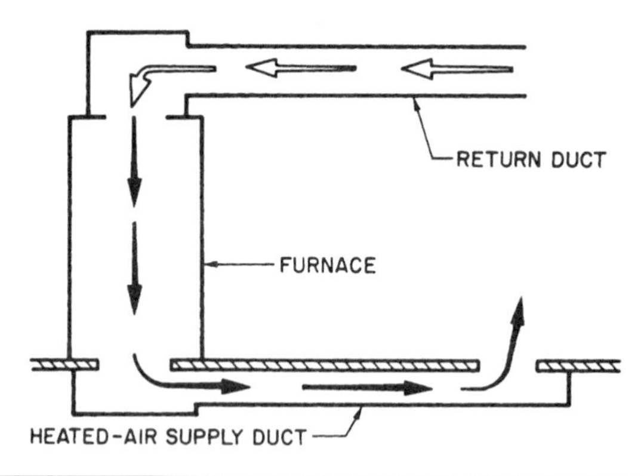

Figure 11-10 Downflow furnaces are commonly found in mobile homes.

Downflow furnaces are commonly found in mobile homes, but may also be used in other residential applications.

Air Handling Systems

Two types of air handling systems found in gas furnaces are the direct-drive motor and the belt-driven blower assembly. In either case, a squirrel cage fan assembly draws air through its center and discharges it through its curved blades located on the outer circumference of the blower. Figure 11-11 shows the two types of blower assemblies found in gas furnaces. In many cases, a multi-speed motor is used in the manufacture of the furnace, but only one speed is used unless the furnace is installed in a split system application, in which the low speed is necessary for the heating mode, and the high speed is necessary for the cooling mode.

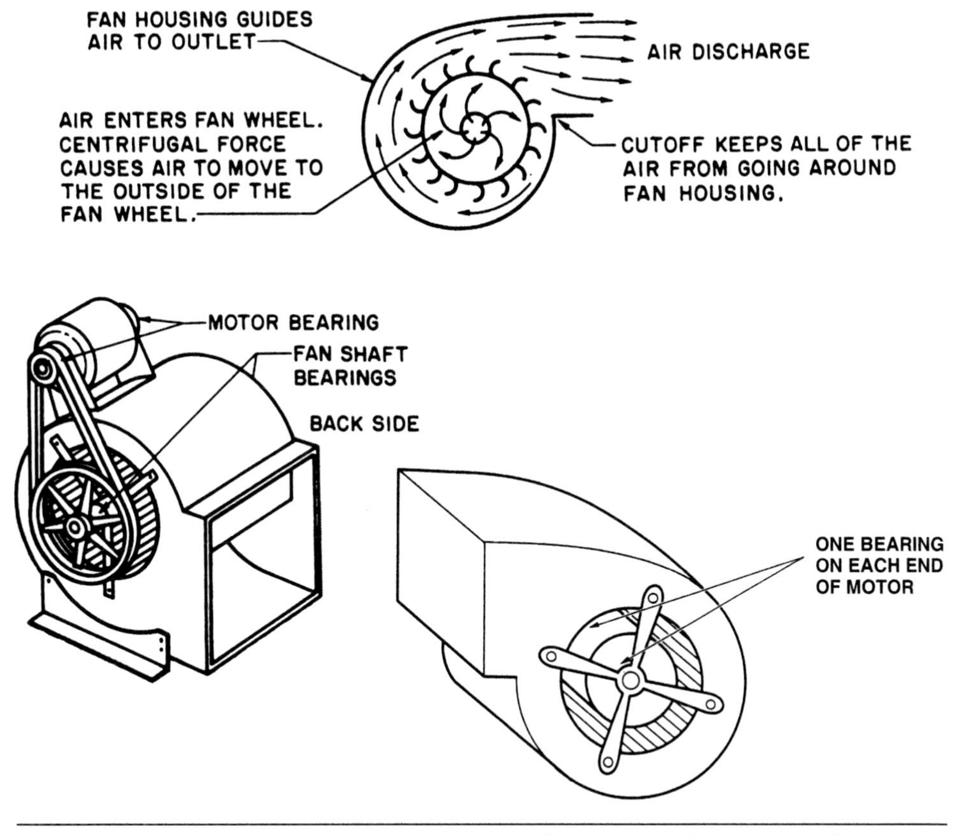

Figure 11-11 Two types of air handling systems found in gas furnaces. Air flow pattern is shown above.

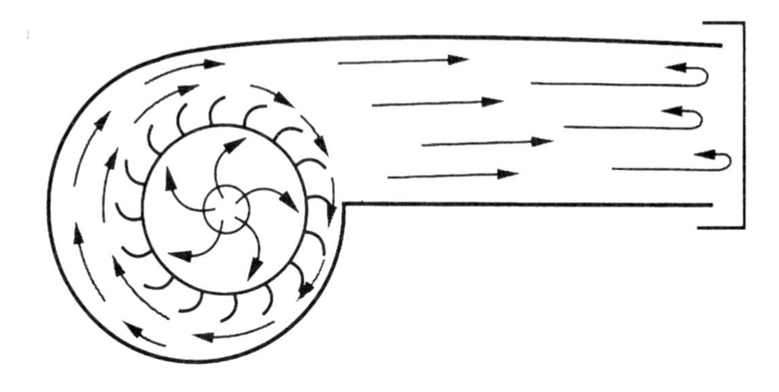

Figure 11-12 When the fan meets too much outlet pressure, the air cannot leave the fan and some of it goes round and round.

TROUBLESHOOTING HINT

If you are called in to check a gas furnace that is providing insufficient heating capacity, make sure there is proper air flow through the furnace. If the air flow volume does not meet design standards, a high limit switch in the furnace could be causing the burner to short cycle, keeping the temperature below the desired setting on the thermostat. A dirty filter (located in the return airstream) could also be the problem, or there could be a problem with the ductwork on the supply side of the system. Figure 11-12 shows what happens to the operation of a squirrel cage blower in the event of a restricted supply duct.

Combustion Systems

When learning to troubleshoot gas furnaces, it is important to have a basic understanding of the combustion system. Figure 11-13 shows the fundamental process through which the byproducts of combustion are discharged through the furnace's vent system.

WARNING

When servicing gas furnaces, always remember that the vent system must be operating properly to prevent dangerous fumes from migrating into the home. If you disconnect a vent or flue box for any reason while servicing a furnace, make sure everything is properly re-installed before leaving the job!

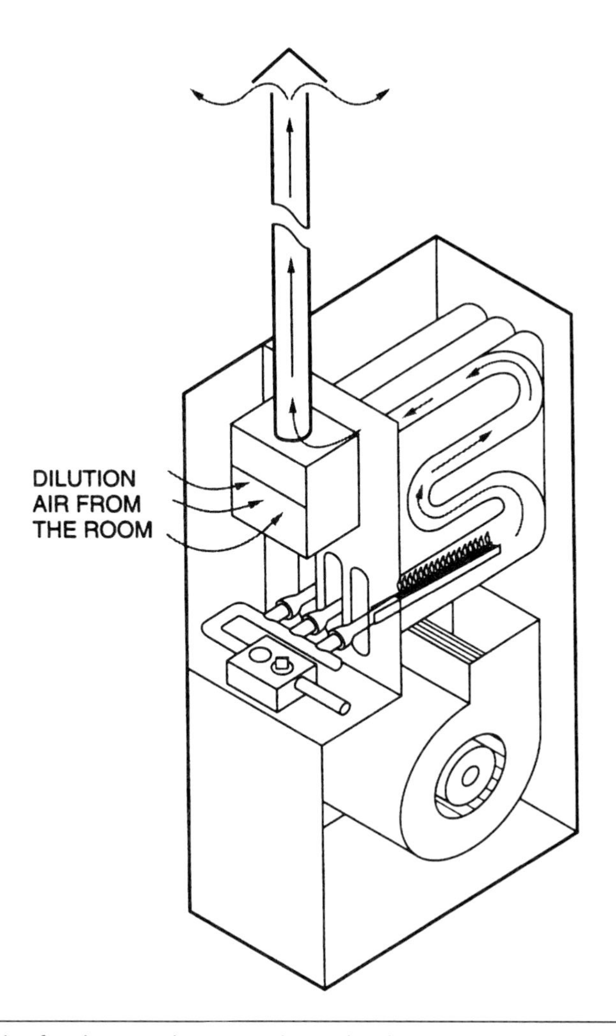

Figure 11-13 The fundamental process through which the byproducts of combustion are discharged through the furnace's vent system.

Ignition Systems

Two methods of ignition in gas furnaces are the standing pilot system and the electric ignition system.

Standing Pilot Systems – The standing pilot system uses what is known as a *thermocouple* to keep the gas valve open and ready to allow gas flow to the main burner on a call for heat by the thermostat. A thermocouple is a device that contains two dissimilar metals, and when heat is applied to the junction point of these two metals, a small amount of voltage is generated.

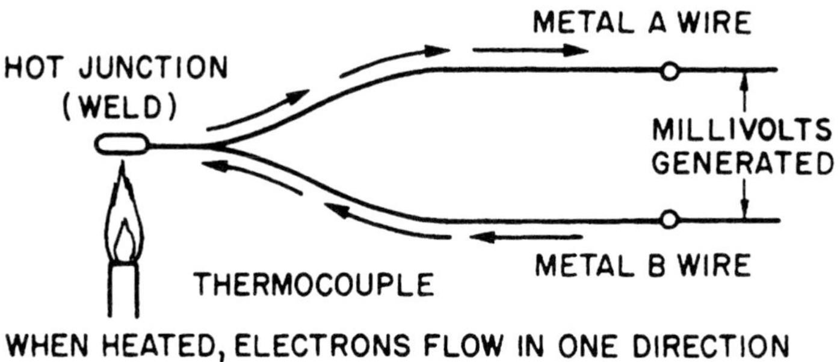

Figure 11-14 The thermocouple is two dissimilar metals with a hot junction and a cold junction. (Courtesy Robertshaw Controls Company)

Figure 11-14 shows the basic theory behind the operation of a thermocouple.

When starting a gas furnace up for the season and first lighting the pilot, the knob on the gas valve is held in place until the heat from the pilot flame is sufficient to generate the voltage needed to hold the coil on the valve open (see Figure 11-15).

In the event that a standing pilot light will not stay lit, the possibilities are that the thermocouple is defective, the gas valve itself is defective, or the pilot flame is not adjusted properly. Some "technicians" use the *parts changing technique* in troubleshooting this problem: first replacing the thermocouple, then testing the furnace operation. If it works, the problem is solved. If it fails to work, they replace the gas valve.

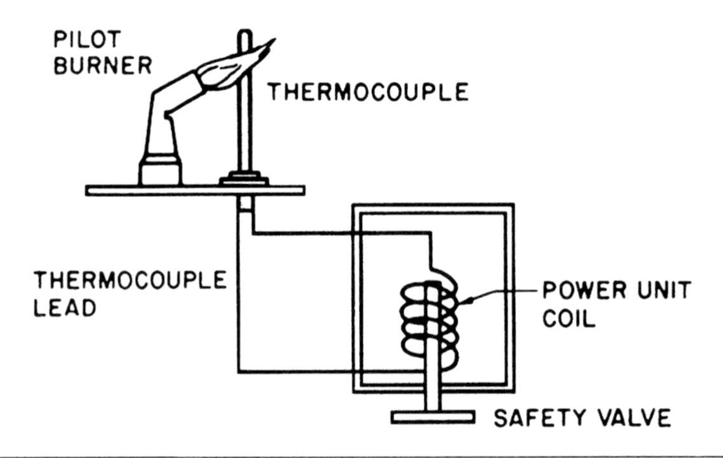

Figure 11-15 The thermocouple holds the valve open for gas to flow.

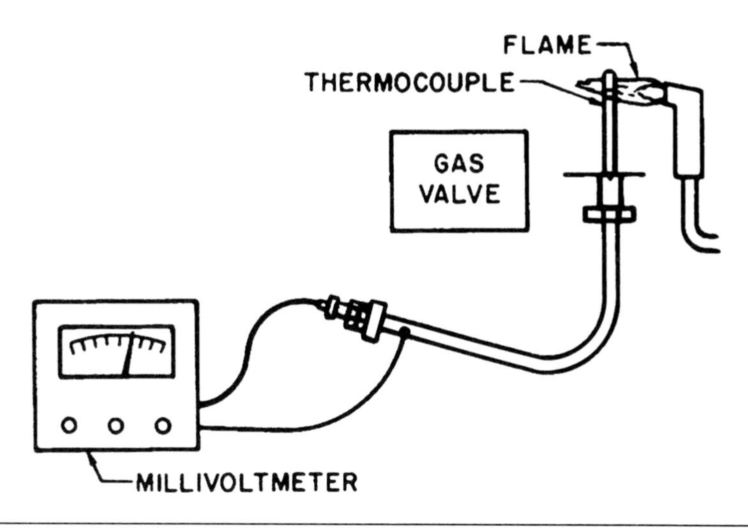

Figure 11-16 Thermocouple operating without a load.

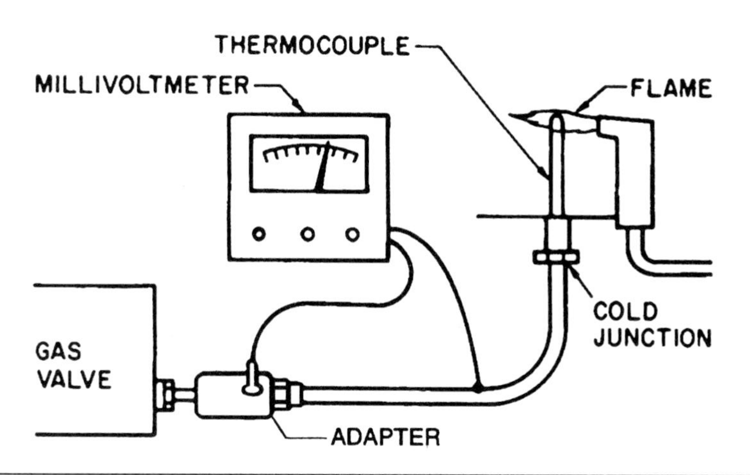

Figure 11-17 Thermocouple operating under a load.

A "no-fail" troubleshooting method for testing a thermocouple system in a gas furnace is to use a millivolt meter and check the operation of the thermocouple under no-load and load conditions. Figures 11-16 and 11-17 outline the sequence for properly troubleshooting a thermocouple system, first under no-load conditions, then under load conditions.

Electronic Ignition Systems – Three types of electronic ignition systems are the glow coil ignition system, the hot surface ignition system, and the spark ignition system.

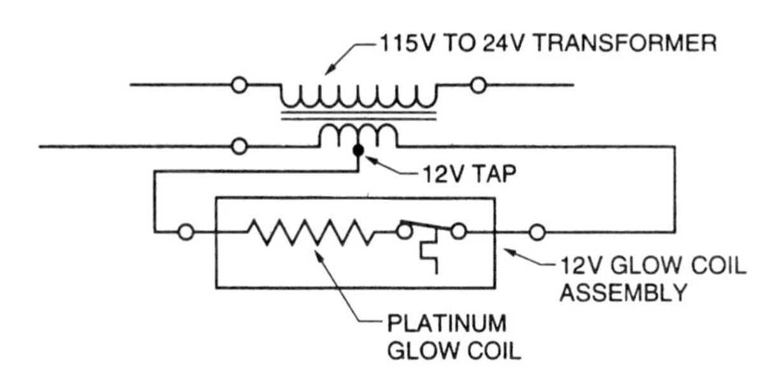

Figure 11-18 Glow coil ignition is accomplished with approximately 12 volts.

A glow coil ignition system uses a small coil of wire to which 12 volts is applied. Either a center tap on a step-down transformer or a resistor wired in series with the glow coil supplies the 12 volts to the coil. In some cases, the varied length of the lead wire to the glow coil provides the correct amount of resistance in the 24-volt control voltage wiring system, dropping half the voltage in the resistive wire so the correct voltage level is delivered to the glow coil. Figure 11-18 shows one method of wiring a glow coil ignition system.

A hot surface ignition system is similar to glow coil operation and will be found on newer units. Older units use glow coil systems. Like a glow coil, a hot surface ignitor glows bright red, causing the gas flow passing nearby to ignite. The hot surface ignition component that glows is made of silicon carbide and is often referred to as a *glow-bar* or a *carborundum coil*. The silicon carbide material is very brittle and can be easily broken if jostled during a furnace servicing procedure. Figure 11-19 shows the components in a hot surface ignition system.

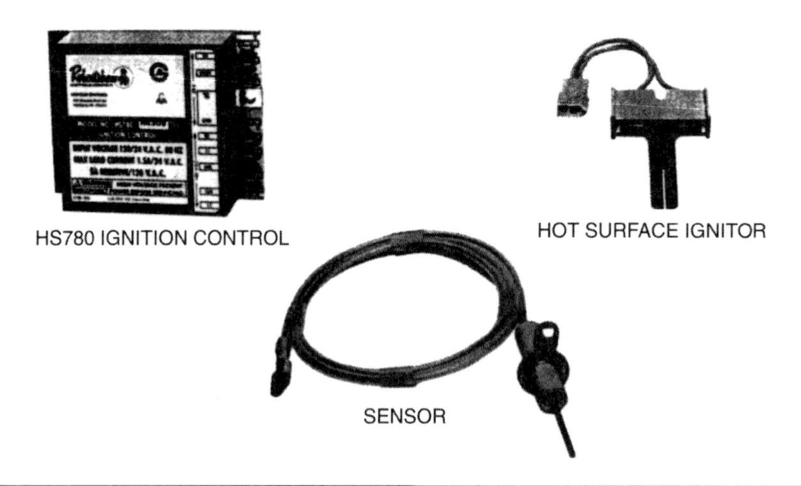

Figure 11-19 Hot surface ignitor. (Courtesy Robertshaw Controls Company)

Figure 11-20 The position of the electrode in a spark ignition system is critical to its operation. (Courtesy Robertshaw Controls Company)

The spark ignition system differs from the glow coil and hot surface ignitor in that it is intermittent, providing an arc rather than a constant glow. On a call for heat from the thermostat, the spark device is energized, providing an arc between its electrode and the pilot bracket, which is fundamentally ground in the electrical system. Figure 11-20 shows the position of the electrode that provides the small gap needed to create the arc.

TROUBLESHOOTING HINT

The spark device on an electronic ignition system should ignite the pilot after about 4 to 6 clicks. If the pilot flame does not ignite and the clicking continues, the problem could be the gap adjustment between the electrode and the pilot bracket assembly. If the electrode is too close to the bracket, the arc is too small to ignite the pilot gas.

Figure 11-21 The type of heating element used in an electric furnace. (By Bill Johnson)

ELECTRIC FURNACES

Electric furnaces use a resistance-type heating element that operates on 240V. When voltage is applied to the element, it glows bright red and provides heat to the air passing through the furnace cabinet. Figure 11-21 shows the type of heater used in an electric furnace.

Depending on the capacity of the furnace, there may be anywhere from two to five elements mounted in the cabinet, as shown in Figure 11-22. An electric furnace can be of the upflow, downflow, or horizontal design. All three styles are shown in Figure 11-23.

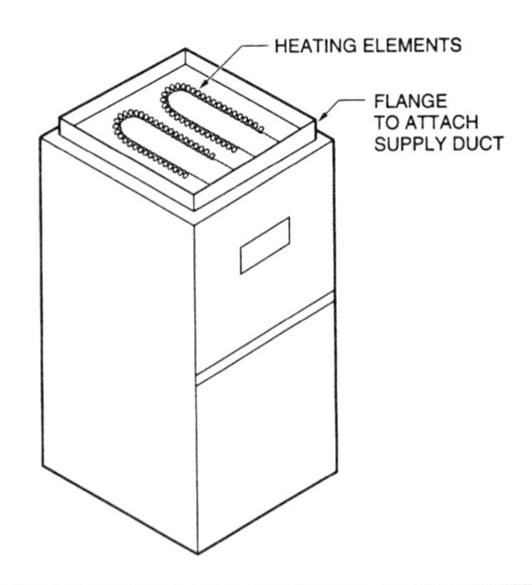

Figure 11-22 In an electric furnace, the heating elements are often located at the top of the cabinet assembly.

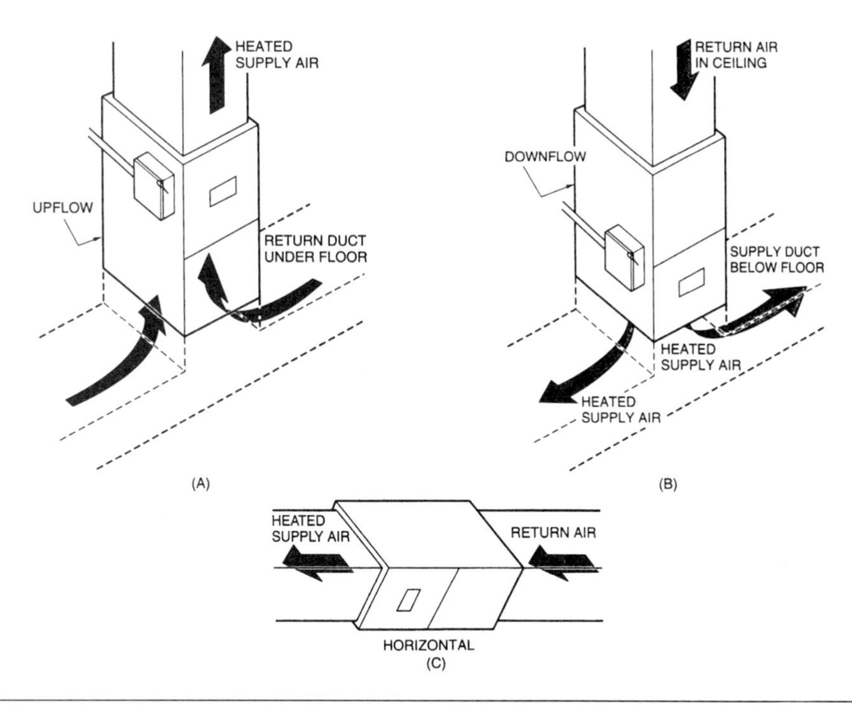

Figure 11-23 (A) Upflow furnace. (B) Downflow furnace. (C) Horizontal furnace.

Troubleshooting an electric furnace is limited to two categories: air flow and electrical. Like any furnace, an electric furnace contains devices that act as safety switches and break the circuit to the heat source in the event of insufficient air flow. Figure 11-24 shows a unit with electric heat.

Figure 11-24 The heating elements are positioned in the furnace cabinet behind the control components. (Courtesy Rheem Air-Conditioning Division)

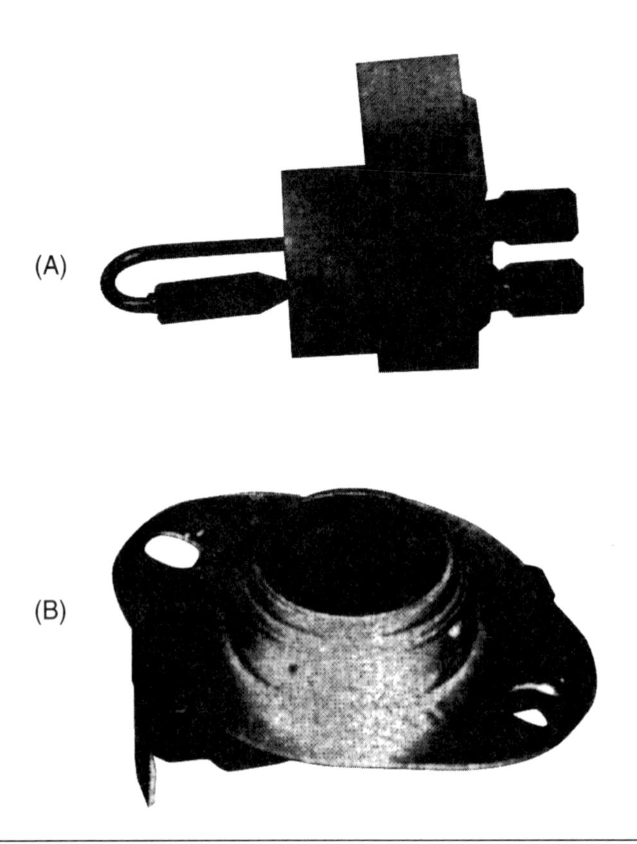

Figure 11-25 The fuse link (A) and the bimetal safety switch (B) are used in an electric furnace. (By Bill Johnson)

Two devices that react to a restricted air flow condition are the fuse link and the bimetal thermodisc. Figure 11-25 shows each type. The bimetal type can be manually reset in the event of a trip. The fuse link type, which consists of a fuse mounted in a ceramic base, must be replaced in order to restore the complete circuit to the heating element.

Both of these safety devices are normally closed, and open only in the event that they are affected by a condition of excessive heat. Safety switches that act as safety devices are mounted in the furnace cabinet in such a way as to sense temperature, and the only portion visible is the point at which the electrical connections are made. Removing one or two screws, then pulling out on the safety assembly, will reveal the sensing section of the device.

In addition to the safety devices described, it is also common to find an automatic-reset temperature sensitive device. This switch operates on

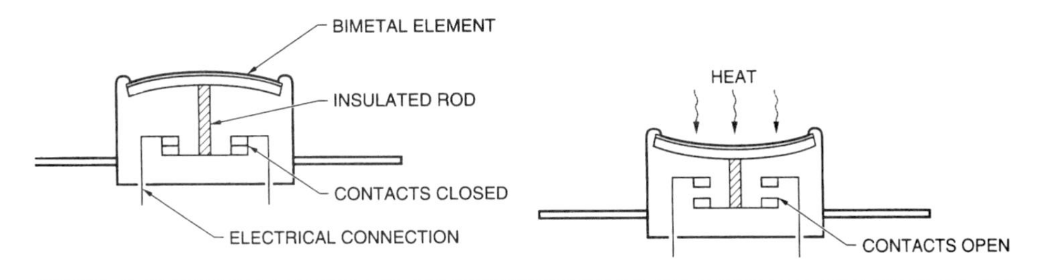

Figure 11-26 The bimetal device reacting to heat.

the same principle as the manual-reset device, but it resets itself after the temperature drops below its trip point. Figure 11-26 shows the method of operation of this device.

Sequencers

Another key component in electric furnaces is the sequencer, a simplified version of which is shown in Figure 11-27.

The method of operation of the sequencer is as follows. When 24 volts is applied to the heater section of the sequencer (H1 and H2), the main switch (contacts M1 and M2) will close after a short time delay, then the auxiliary switch (contacts A1 and A2) will close in a delay sequence a short time later. Some sequencers may contain up to four auxiliary switches, meaning that the device can handle up to five separate loads in a sequence operation.

The purpose of the sequencer in an electric furnace is to provide a "step" system for energizing the heating elements. If all the elements in a furnace were energized at the same time, it would create a momentary condition of excessive voltage drop.

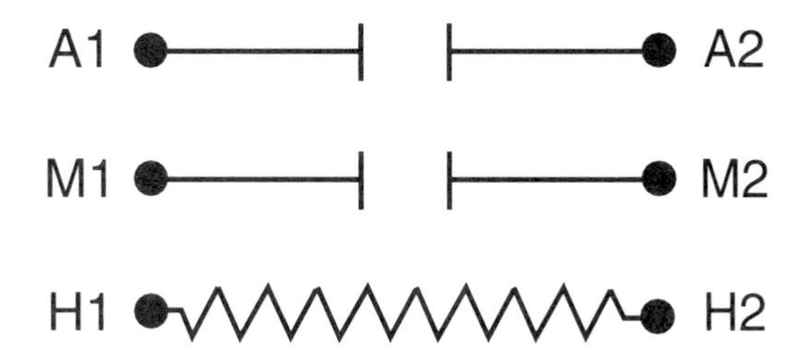

Figure 11-27 A common type of sequencer contains two separate switching circuits that are controlled by the heater section of the device.

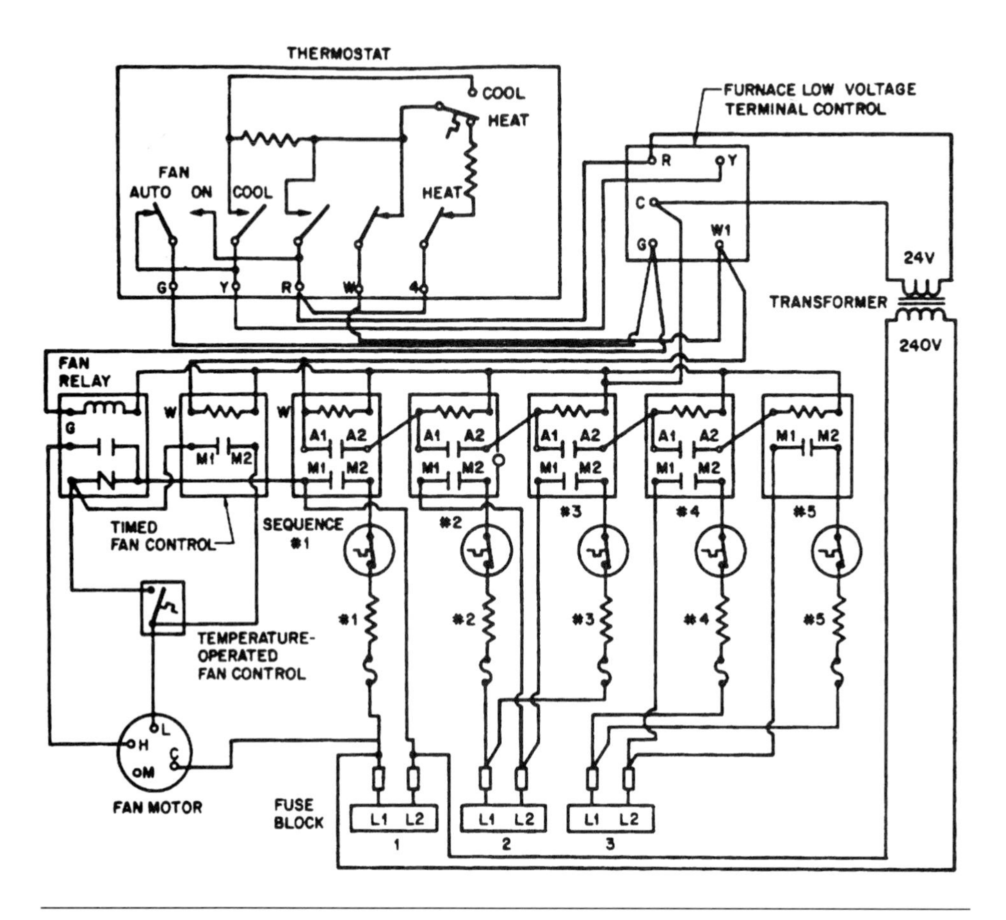

Figure 11-28 A pictorial diagram of the components in an electric furnace.

An important thing to keep in mind about sequencers is the difference between the way they are depicted in pictorial and schematic diagrams. In Figure 11-28, for example, the sequencers are shown as a complete component.

When shown schematically, the sequencer circuit function is split between the high voltage and low voltage circuits. Figure 11-29 shows a schematic layout of an electric furnace that uses five heating elements, and consequently, five sequencers.

To trace the sequencer operation on a schematic, first establish that on a call for heat from the thermostat, 24 volts would be applied to the heater section of the #1 sequencer. (Identified as SEQ #1 and located right below the timed fan control heater section on the low voltage side of the diagram.)

Remember that the first assignment the sequencer must accomplish is the closing of contacts M1 and M2, the main switching assembly.

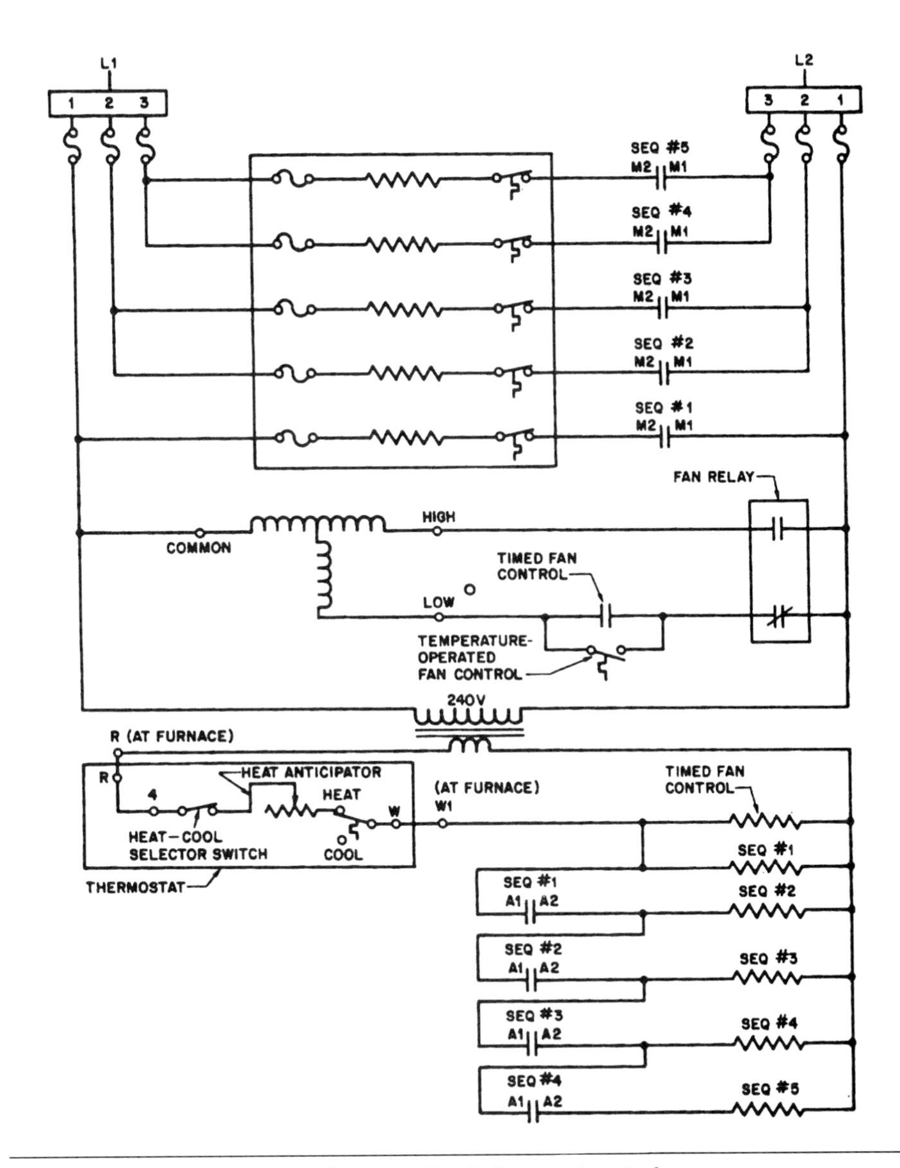

Figure 11-29 A five-element heating circuit for an electric furnace.

On this schematic, these switching points are wired in series with a heating element shown on the high voltage side of the diagram and identified as SEQ #1, M1 and M2. Once these switch contacts close, a 240-volt circuit is completed to the heating element, provided that the fuse link and bimetal switches also shown wired in series with the heating element are in their normal position.

The second assignment of the sequencer is to close switching contacts A1 and A2. In Figure 11-29, these contacts on SEQ #1 are shown wired in series with the heater section on the second sequencer. This brings us back to the low voltage side of the diagram.

Identifying the circuit to the heater section of SEQ #2 returns us to the high voltage section of the diagram to determine the function of the main switching contacts of the sequencer (M1 and M2 of SEQ #2). This illustrates the pattern of switching from the high and low voltage sections of the diagram to trace the function of the sequencer, even though it appears physically as one component.

For a view of how these components may be found in an actual electric furnace, see Figure 11-30.

Figure 11-30 The internal components of an electric furnace. (Courtesy Rheem Air-Conditioning Division)

PROBLEM 7 AN ELECTRIC FURNACE IS PUTTING OUT SOME HEAT, BUT NOT AS MUCH AS IT SHOULD

Upon arrival, you find the following conditions:

1. When the thermostat is turned up, the blower motor starts after a slight delay and the air being delivered by the unit is cool, not warm.
2. After a longer delay, the air temperature from the furnace is warmer, but still not up to design parameters.

To troubleshoot this problem, proceed as follows:

STEP ONE: With the furnace in full operation for five minutes, test for 240 volts applied to each element. (Refer to Figure 11-31.) At this point, you find that voltage is applied to elements 2, 3, 4, and 5, but not to element 1. Proceed to step two.

STEP TWO: To isolate the problem, leave one lead of a voltmeter on the L1 side of the connection at the heating element and trace backward with the other lead of the meter through the bimetal switch and the sequencer switching contacts M1 and M2. There is no voltage at M2, but there is voltage at M1. Proceed to step three.

STEP THREE: Switching to the 24-volt circuit, measure the voltage applied to the heater section of SEQ #1, as shown in Figure 11-32. The voltmeter reads 24 volts.

Which component is at fault and what is the specific failure of the component?

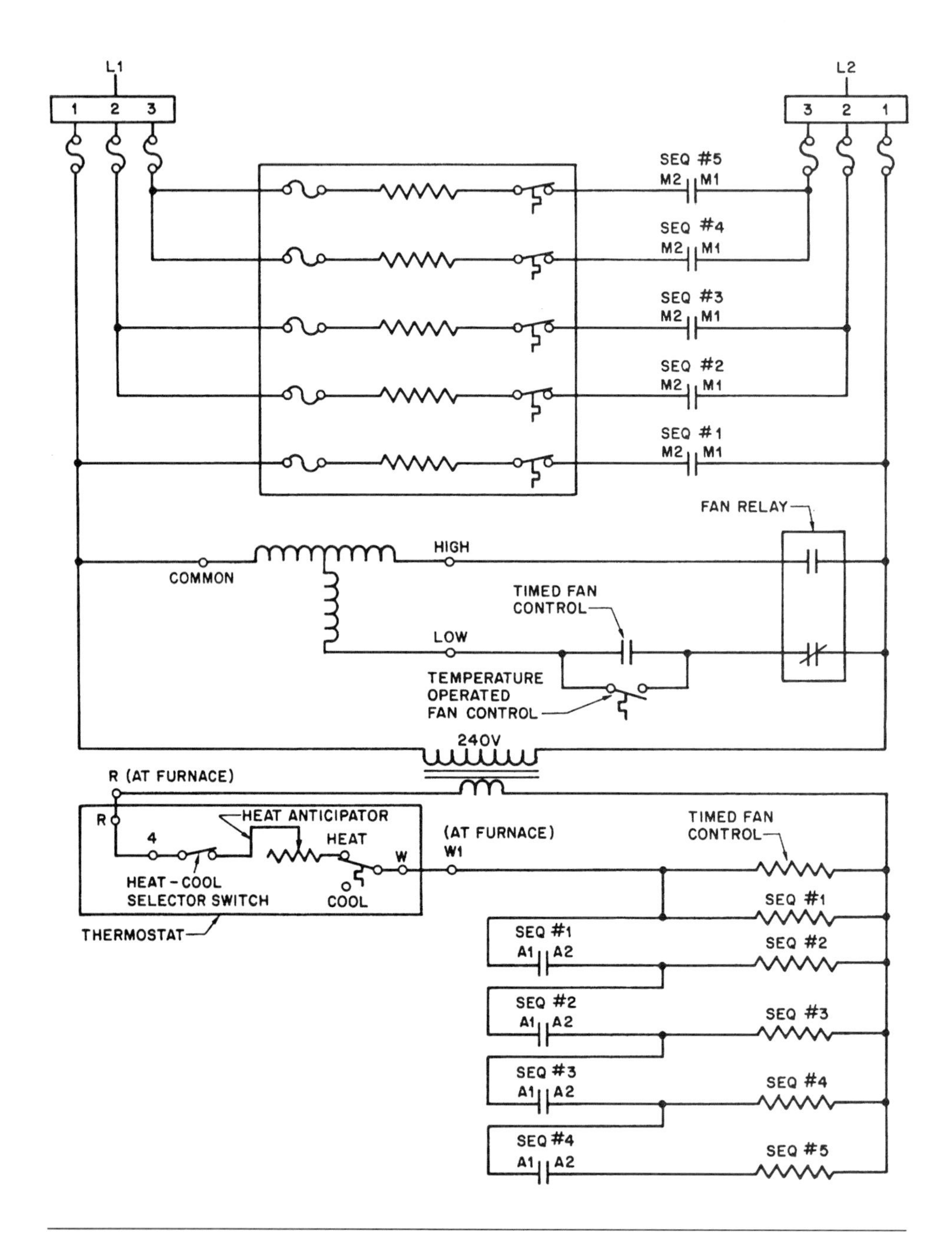

Figure 11-31 Sequencer control circuit wiring.

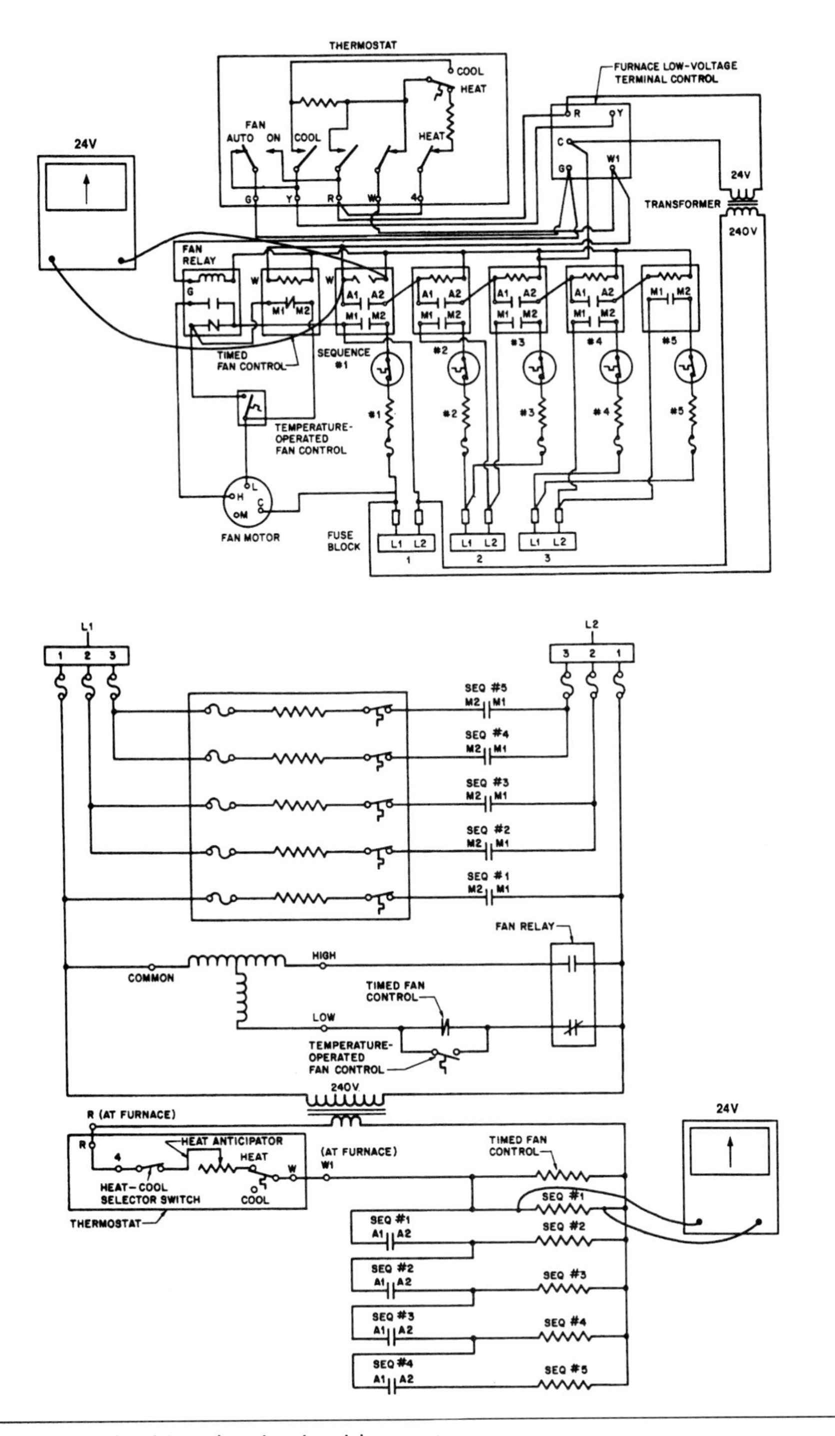

Figure 11-32 Checking the circuit with a meter.

PROBLEM 8 *A GAS FURNACE IS NOT HEATING AT ALL*

Upon arrival, you find the following conditions:

1. The temperature in the house is 60°F.
2. The thermostat is set to 80°F.
3. The furnace is not running at all (no fan, no operation of the gas burner).

To troubleshoot this problem, proceed as follows:

STEP ONE: Remove the cord from the 120-volt receptacle and test it with a voltmeter. Voltage is read in the receptacle. Proceed to step two.

STEP TWO: After replacing the power cord in the receptacle, remove the furnace door and the cover from the unit's printed circuit board (shown in Figure 11-33). Connect a voltmeter to the terminals identified as PR-1 and PR-2. Proceed to step three.

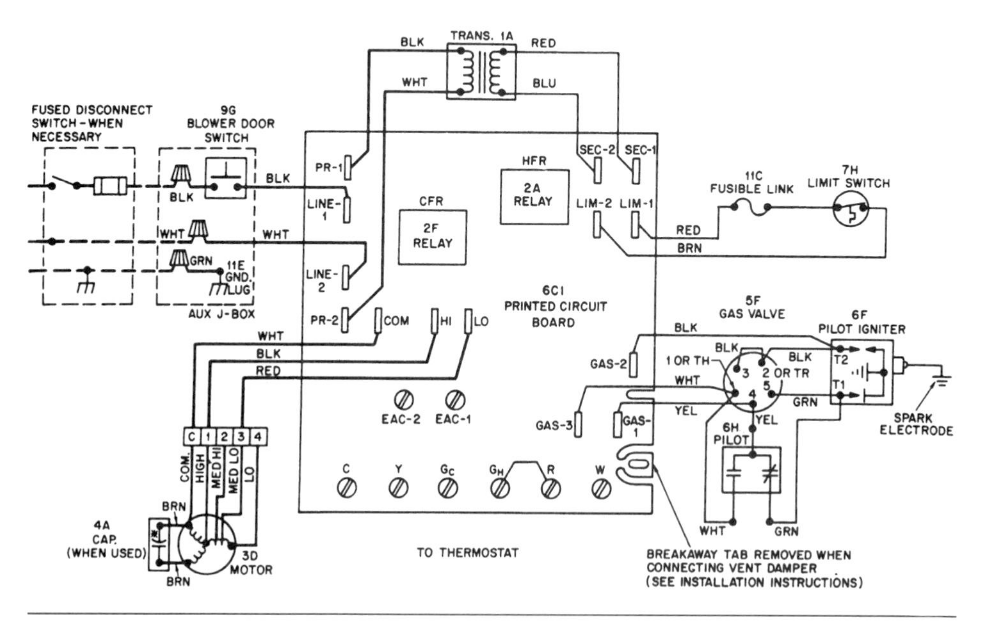

Figure 11-33 Check for power at PR-1 and PR-2. (Courtesy Bryant Heating & Cooling Products)

STEP THREE: When you push on the blower door switch button that would normally be depressed when the furnace door is in place, the voltage reading is 120 volts. Proceed to step four.

STEP FOUR: Switch the leads of the meter to the terminals identified as SEC-1 and SEC-2. No voltage is read when you press on the furnace blower switch.

Which component has failed?

PROBLEM 9 *A GAS FURNACE BURNER IGNITES, BUT GOES OFF BEFORE THE AIR HANDLING SYSTEM BEGINS TO OPERATE*

Upon arrival, you find the following conditions:

1. The pilot on the furnace is lit.
2. The temperature in the house is 60°F.
3. The thermostat is set to 80°F.
4. A short time after arrival, the furnace main burner ignites, stays on for approximately 60 seconds, then shuts off. The pilot remains lit. This cycle repeats in approximately 2 minutes.

To troubleshoot this problem, proceed as follows:

STEP ONE: (Refer to Figure 11-34.) Place a voltmeter across the fan switch section of the fan/limit switch assembly. The voltage reading is 120 volts. Proceed to step two.

STEP TWO: Wait for the burner to ignite. After the burner comes on, you continue to read 120 volts at the switch contacts. Proceed to step three.

STEP THREE: The burner shuts off after about 60 seconds. Switch your voltmeter leads across the limit switch section of the fan/limit switch assembly. The voltmeter reads 120 volts.

Which component must be replaced in order to accomplish the repair?

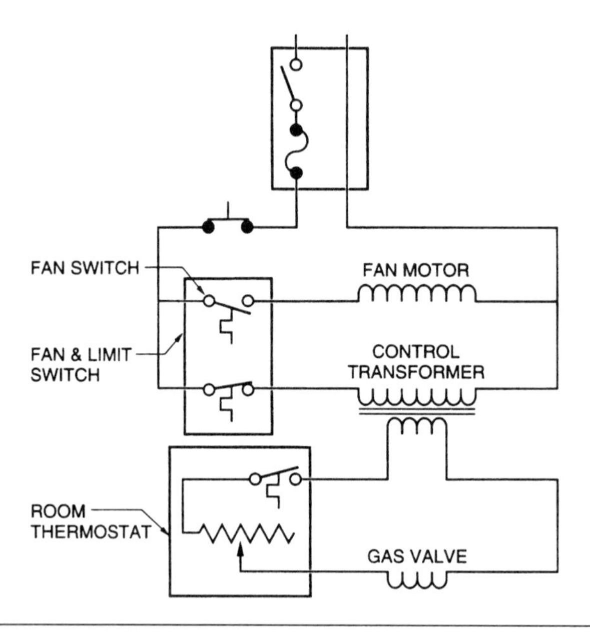

Figure 11-34 Schematic diagram for a gas furnace.

PROBLEM 10 *A SMALL PACKAGED UNIT AIR CONDITIONER IS NOT COOLING*

Upon arrival, you find the following conditions:

1. The temperature in the house is 85°F.
2. The thermostat is in the AUTO position and is set at 65°F.
3. The indoor fan is running and the unit is blowing warm air.

To troubleshoot this problem, proceed as follows:

STEP ONE: Remove the unit access panel and test for voltage at terminals T1 and T2 (refer to Figure 11-35). The reading is 240 volts. Proceed to step two.

STEP TWO: The dome of the compressor feels warm. In a moment, the compressor hums for approximately 2 seconds, then kicks off. Proceed to step three.

STEP THREE: Turn off the disconnect switch, remove the wires from the run capacitor, and test it with an analog ohmmeter. The needle on the meter rises, then drops to zero.

What is one possible repair that will solve the problem?

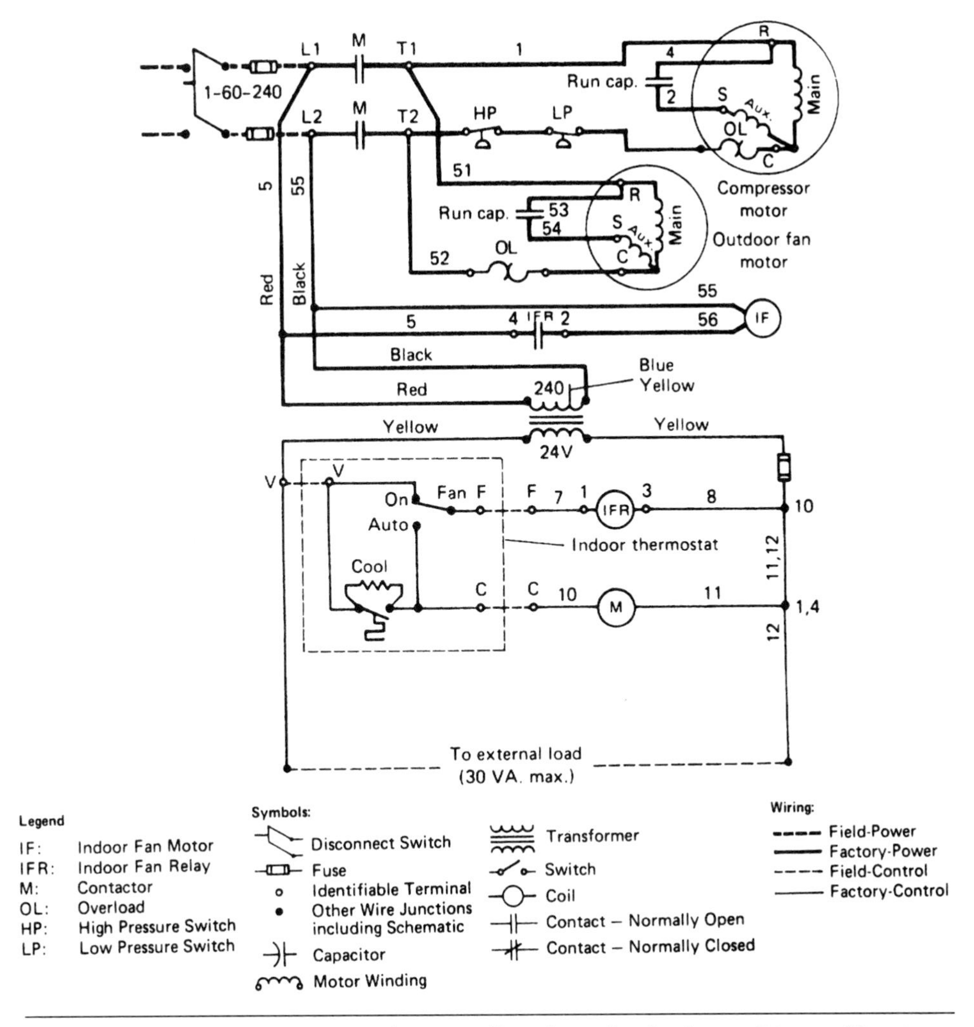

Figure 11-35 Schematic diagram for a small packaged unit air conditioner. (Courtesy York International, Inc.)

PROBLEM 11 *A CENTRAL AIR CONDITIONING SYSTEM IS RUNNING BUT NOT COOLING PROPERLY*

Upon arrival, you find the following conditions:

1. The indoor temperature is 80°F.
2. The thermostat is set at 70°F. The fan switch is in the AUTO position.

To isolate this problem, proceed as follows:

STEP ONE: Remove the access panel to the compressor and electrical section. The compressor is running and the suction line is frosted. Proceed to step two.

STEP TWO: Remove the cover on the filter slot on the return air duct. There is no filter in the unit. Proceed to step three.

STEP THREE: Inside the residence, test with a velocimeter at a discharge register to determine the air flow. Record the volume of air. Proceed to step four.

STEP FOUR: To complete the evaluation, remove the access panel on the roof that exposes the indoor fan motor. Proceed to step five.

STEP FIVE: Back inside the house, test the same discharge register. The volume of air is significantly increased.

What must you do in order to restore this unit to full operating capacity?

PROBLEM 12 *A SPLIT SYSTEM A/C UNIT IS NOT COOLING*

Upon arrival, you find the following conditions:

1. The indoor fan motor is running.
2. The temperature inside the house is 85°F.
3. The thermostat is set at 65°F. The fan switch is in the AUTO position.

To troubleshoot this problem, proceed as follows:

STEP ONE: The condensing unit is quiet. Proceed to step two.

STEP TWO: Remove the access panel and use a voltmeter to test at L1 and L2. (Refer to Figure 11-36.) The applied voltage is 240 volts. Proceed to step three.

STEP THREE: Test the coil of the contactor. The applied voltage is 24 volts. Proceed to step four.

STEP FOUR: Test terminals T1 and T2 of the contactor. There is no voltage reading at these points.

Which component has failed?

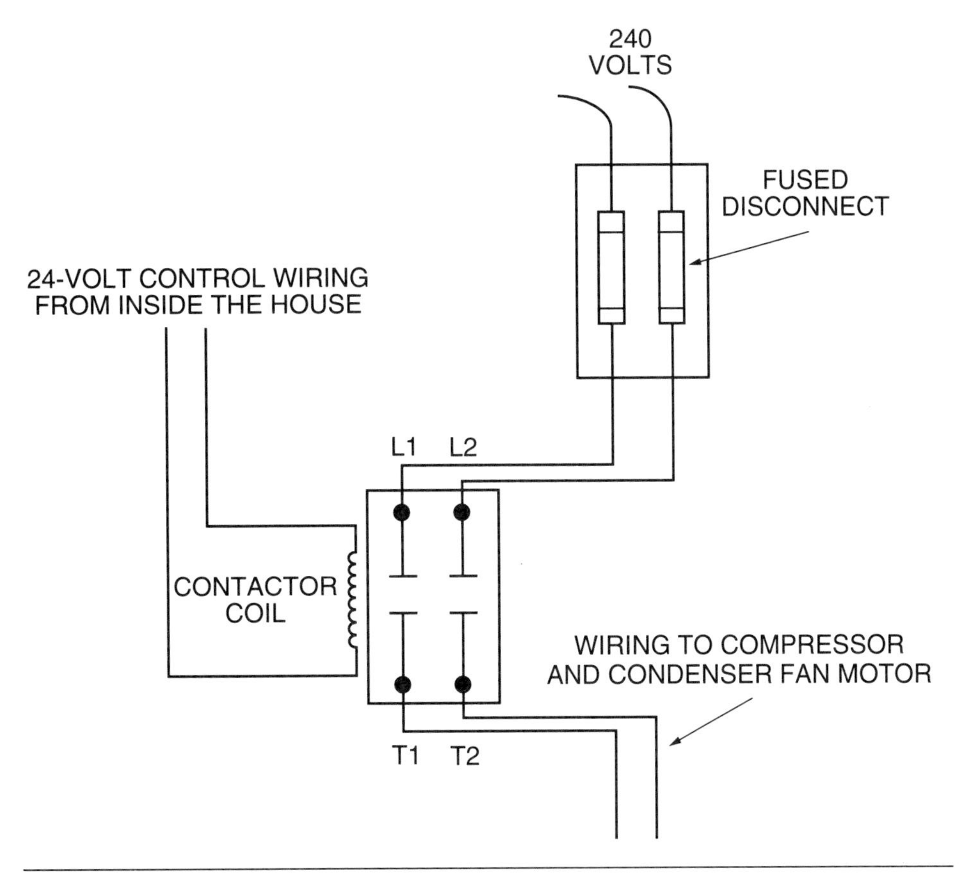

Figure 11-36 Wiring for a split system condensing unit.

PROBLEM 13 *A HEAT PUMP UNIT IS NOT COOLING AND NOT RUNNING*

Upon arrival, you find the following conditions:

1. The temperature inside the house is 85°F.
2. The thermostat is set at 70°F. The fan switch is set in the AUTO position.
3. The unit is not running and the indoor fan is not blowing.

To troubleshoot this problem, proceed as follows:

STEP ONE: Refer to Figure 11-37. At the unit, test for applied voltage at L1 and L2. The voltage is 240 volts. Proceed to step two.

STEP TWO: Test the secondary side of the transformer. The voltage reading is 24 volts. Proceed to step three.

STEP THREE: Test for 24 volts at the CR coil. No voltage is read. Proceed to step four.

STEP FOUR: Turn off the disconnect switch. Inside, disconnect the thermostat wiring and tie R, Y, and G together. Outside, turn the disconnect back on. The unit runs.

Which component must be replaced?

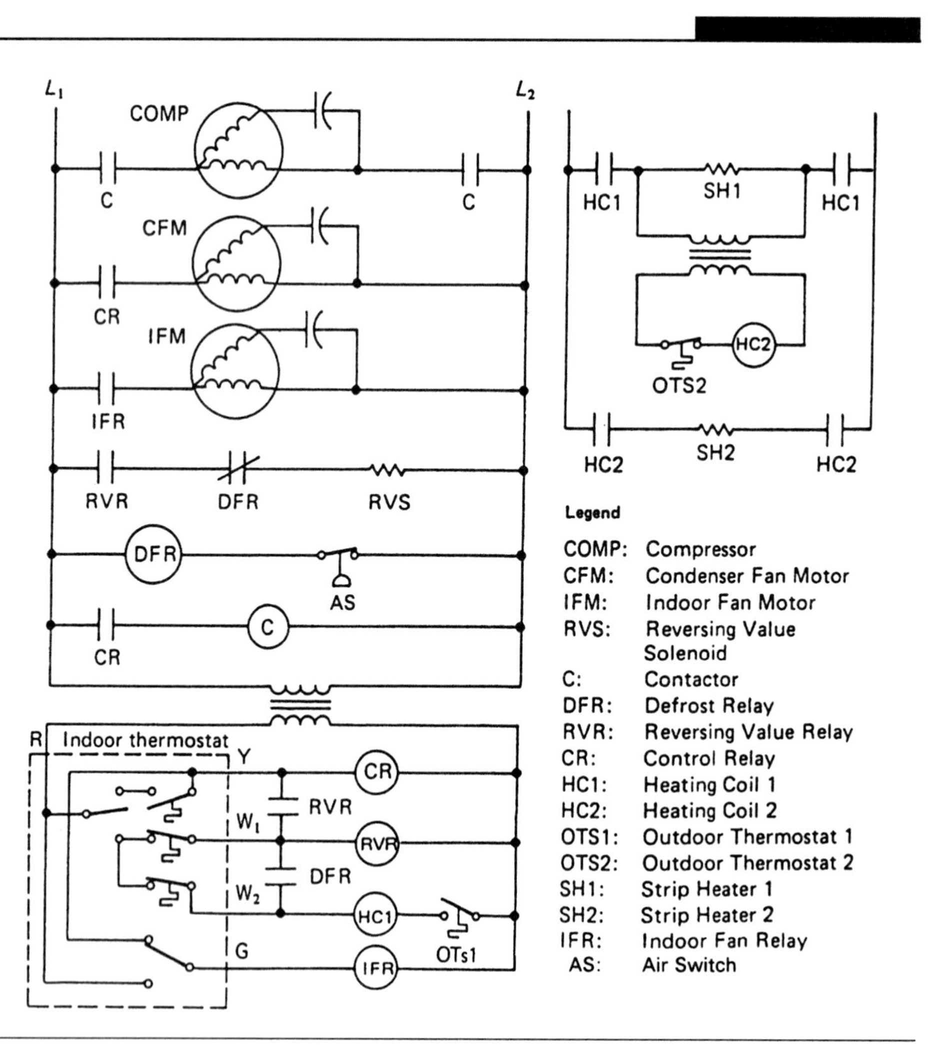

Figure 11-37 Schematic diagram of a heat pump using a two-stage heating and one-stage cooling thermostat to control the unit.

PROBLEM 14 *A COMPRESSOR ON A PACKAGED A/C UNIT RUNS ONLY A SHORT TIME BEFORE KICKING OFF ON THE INTERNAL OVERLOAD*

Upon arrival, you find the following conditions:

1. The temperature in the house is 85°F.
2. The thermostat is set at 75°F. The fan switch is in the AUTO position.
3. The indoor fan is running.

To troubleshoot this problem, proceed as follows:

STEP ONE: Outside, you find that the condenser fan motor is running, and the compressor tries to start, but only runs for approximately three seconds, then kicks off. Proceed to step two.

STEP TWO: Remove the access cover to the electrical components. The compressor is wired as a capacitor start, capacitor run (CSCR) motor, using a potential relay as the start device. (See Figure 11-38.) Proceed to step three.

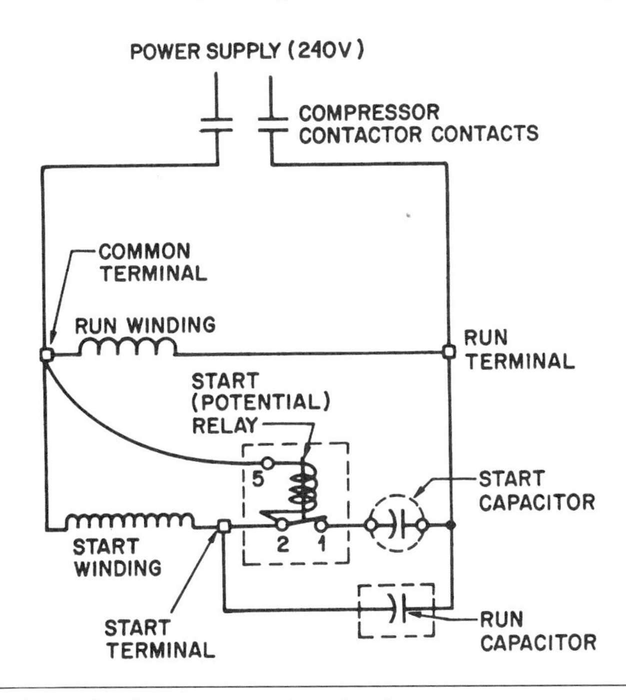

Figure 11-38 Schematic of a motor starting circuit.

STEP THREE: Disconnect the power supply, then disconnect the wiring from the start capacitor. Test the capacitor with an ohmmeter. The meter reacts by rising, then falling back to zero. Proceed to step four.

STEP FOUR: Disconnect the wiring from the run capacitor and test the capacitor with an ohmmeter. The meter reacts by rising, then falling back to zero. Proceed to step five.

STEP FIVE: Restore the wiring and power. Use a voltmeter to test the wiring connections at common and start on the compressor the next time the compressor tries to start. (See Figure 11-39.) The voltage reading is 0 volts.

Which component must be replaced?

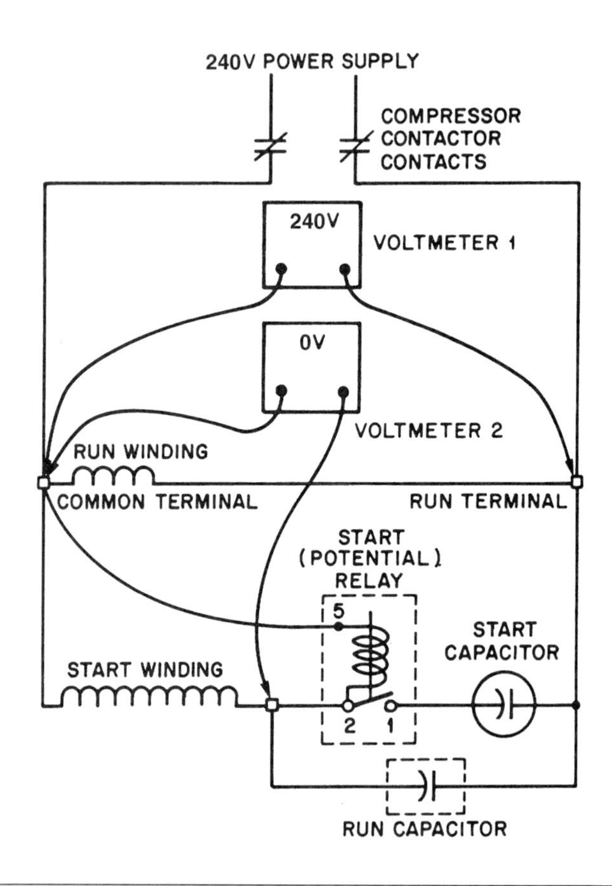

Figure 11-39 Testing the relay terminals.

PROBLEM 15 AN INDOOR FAN ON A PACKAGED GAS UNIT WILL NOT RUN IN THE COOLING MODE

Upon arrival, you find the following conditions:

1. The temperature in the house is 85°F.
2. The customer has turned the unit off, awaiting your arrival.
3. The thermostat is set at 68°F. The fan switch is in the AUTO position.

To troubleshoot this problem, proceed as follows:

STEP ONE: Turn the fan switch from the AUTO position to the ON position. The indoor fan motor does not start. Proceed to step two.

STEP TWO: At the unit, test for 24 volts at the IFM relay coil. The test shows positive. As a follow-up, test for voltage at the common and high connections of the IFM (see Figure 11-40). The voltage reading is 240 volts.

Which component must be replaced?

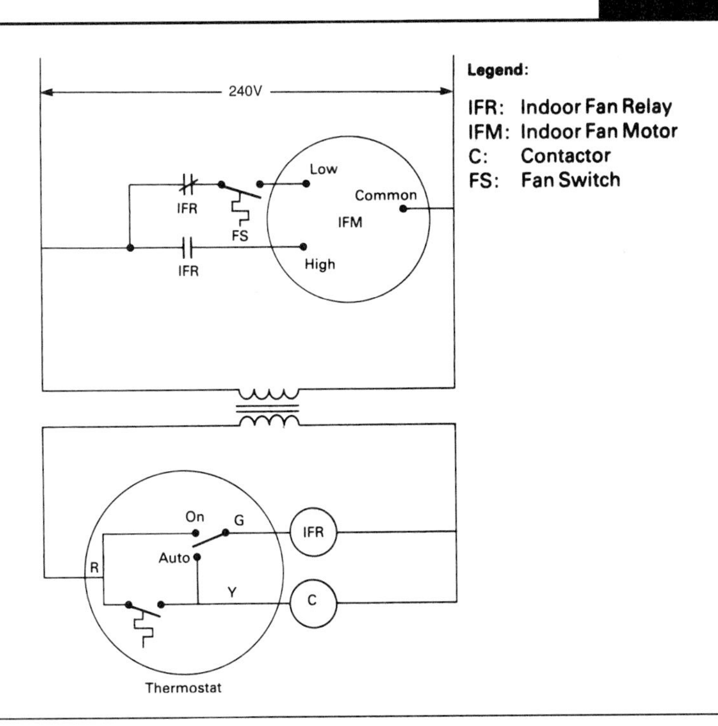

Figure 11-40 Indoor fan motor wiring diagram.

PROBLEM 16 *A PACKAGED HEAT PUMP UNIT IS NOT COOLING PROPERLY*

Upon arrival, you find the following conditions:

1. The temperature inside the house is 85°F.
2. The customer has turned the unit off, awaiting your arrival.

To troubleshoot this problem, proceed as follows:

STEP ONE: Turn the thermostat to the cooling mode, set the fan switch in the AUTO position, and set the thermostat at 75°F to initiate a run cycle. Proceed to step two.

STEP TWO: After a few moments, you confirm the customer's information that the unit seems to be blowing warm air rather than cool air. Proceed to step three.

STEP THREE: At the unit, check terminal connections 1 and 3 (refer to Figure 11-41). The voltage reading is 24 volts. Proceed to step four.

STEP FOUR: As a follow-up check, test for voltage at terminals 2 and 4 of the heat/cool relay. The meter reads 240 volts.

Which component has to be replaced?

PROBLEM 17 *AN A/C UNIT IS NOT COOLING PROPERLY*

Upon arrival, you find the following conditions:

1. The outdoor temperature is 90°F.
2. The indoor fan is running and is blowing warm air.
3. The thermostat is set in the cooling mode with the fan switch in the AUTO position. The thermostat is set at 65°F.
4. The indoor temperature is near 90°F.

To troubleshoot this problem, proceed as follows:

STEP ONE: Check the unit. The condenser fan motor is running. Remove the compressor compartment access cover. The compressor is not running. Proceed to step two.

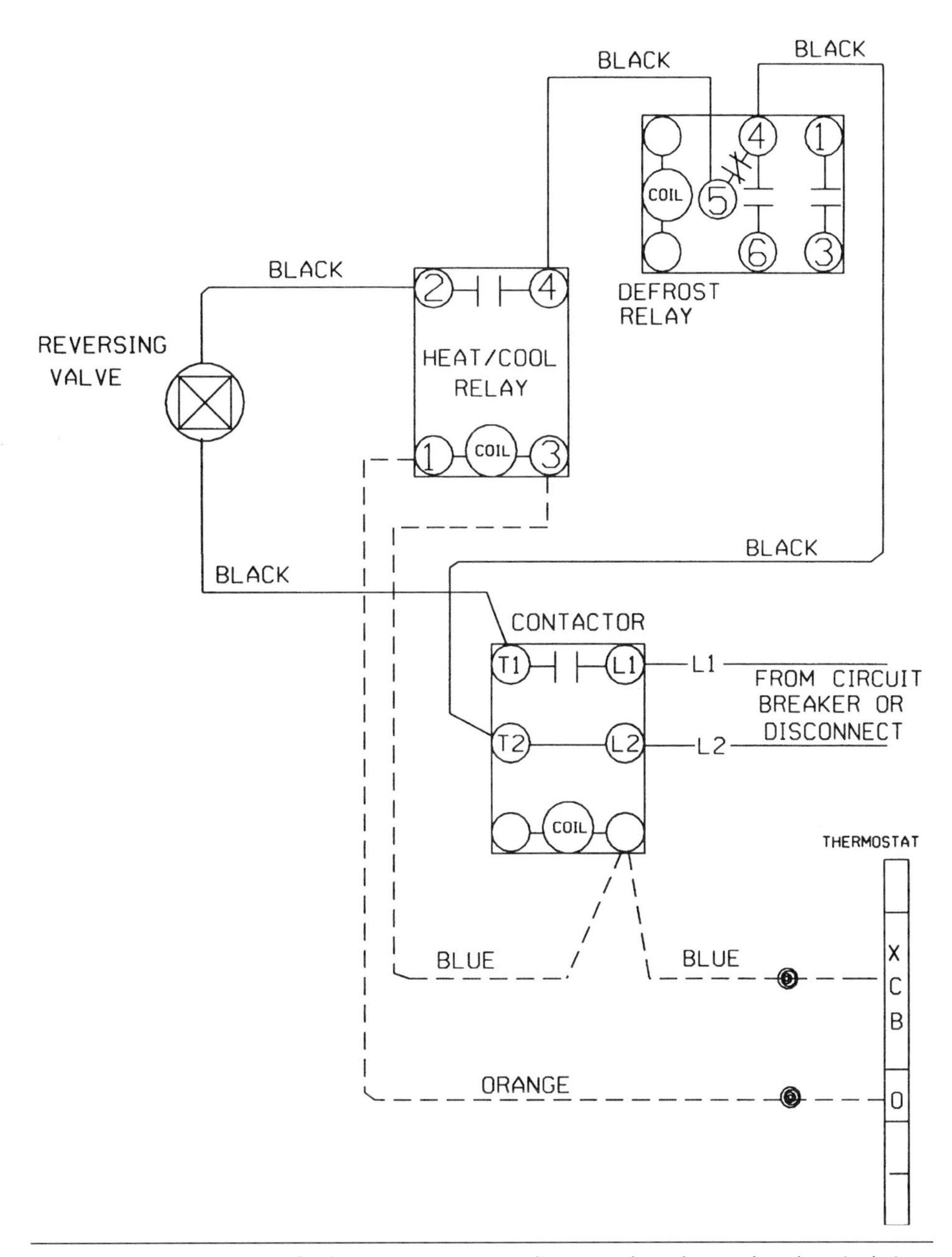

Figure 11-41 A section of a heat pump wiring diagram that shows the electrical circuits for the heating and cooling modes.

STEP TWO: While checking the wiring diagram (refer to Figure 11-42) the compressor starts, runs for about 45 seconds, then shuts down. After about 30 seconds, the cycle repeats. Proceed to step three.

STEP THREE: Connect your gauges to the sealed system, then wait for the next run cycle. (This unit uses HCFC-22 refrigerant.) The low side pressure drops to 20 psig after the compressor starts and runs, then shuts down. While the compressor is off, the low side pressure rises to 55 psig and the compressor cycles again.

What is the problem and what steps must be taken to repair the unit?

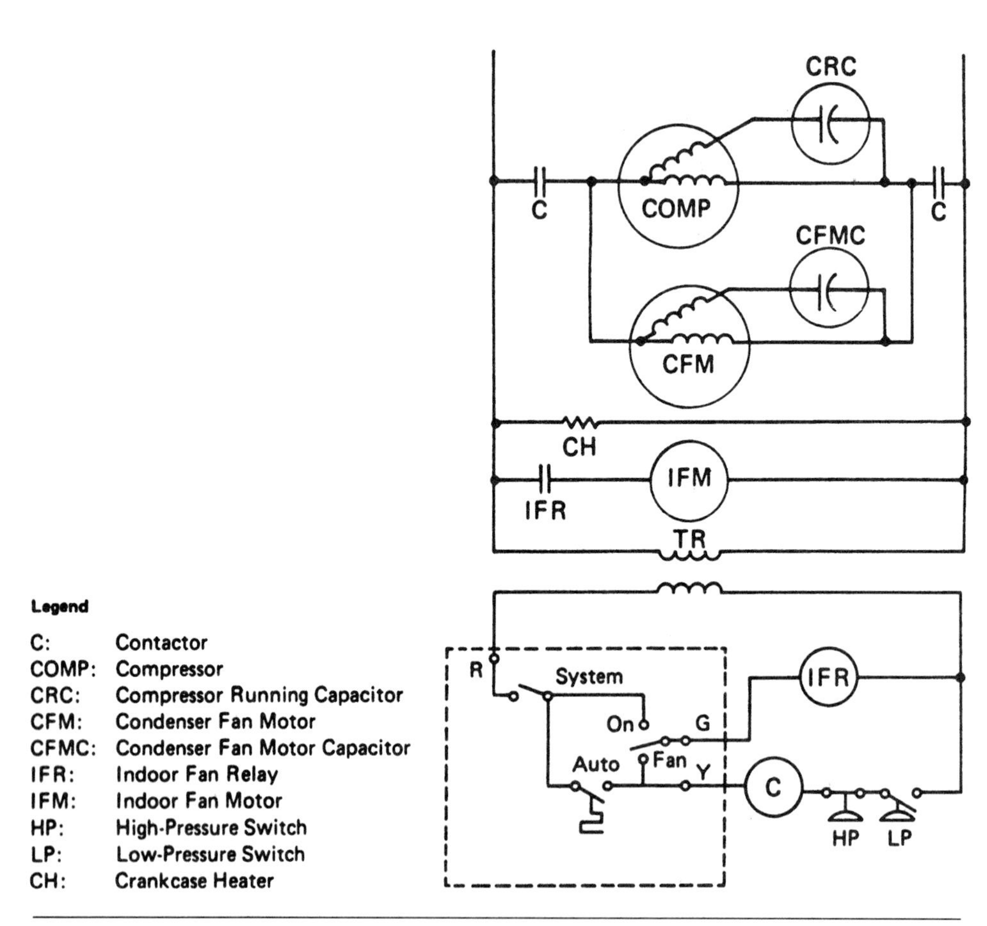

Figure 11-42 A/C unit simplified wiring diagram.

PROBLEM 18 A COMPRESSOR ON A 3-TON PACKAGED UNIT IS NOT STARTING

Upon arrival, you find the following conditions:

1. The thermostat is set in the cooling mode and the fan switch is set in the AUTO position.
2. The indoor fan motor is blowing warm air.

To troubleshoot this problem, proceed as follows:

STEP ONE: Refer to Figure 11-43. When checking the unit, you find that the condenser fan motor is running. Proceed to step two.

STEP TWO: No voltage is applied at terminals C and R or at terminals C and S on the compressor.

What is one possible problem?

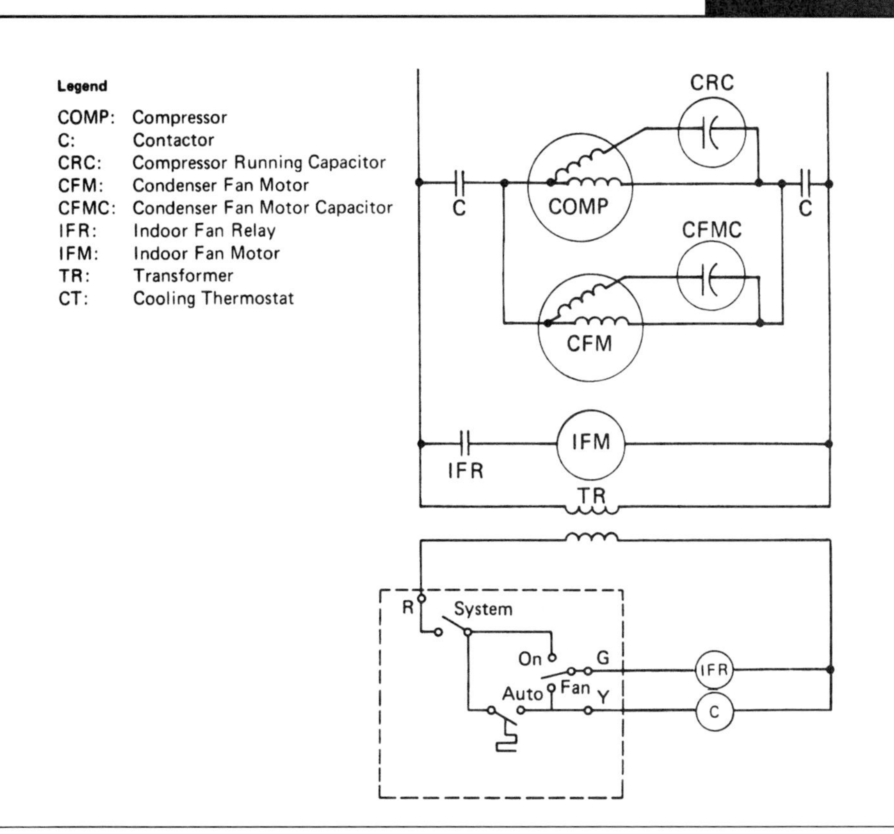

Figure 11-43 A/C unit simplified wiring diagram.

PROBLEM 19 ALL COMPONENTS ON AN A/C UNIT ARE OPERATING, BUT THE UNIT IS BLOWING WARM AIR

Upon arrival, you find the following conditions:

1. The ambient temperature is 90°F.
2. The compressor, condenser fan, and indoor fan are running.

To troubleshoot this problem, proceed as follows:

STEP ONE: Connect your gauges to the sealed system. (This unit uses HCFC-22 refrigerant.) The low side pressure reading is 30 psig and the high side pressure reading is 145 psig.

STEP TWO: The unit nameplate lists the FLA of this unit as 25 amps. Check it with an ammeter. It reads 12.5 amps.

What is the problem and what steps must be taken to repair the unit?

PROBLEM 20 A FUSE BLOWS IMMEDIATELY AND REPEATEDLY WHEN AN A/C UNIT IS SET TO START IN THE COOLING MODE

Upon arrival, you find the following conditions:

1. The thermostat is set in the cooling mode, the fan switch is set to AUTO, and the thermostat set point is 75°F.
2. The indoor temperature is 85°F.
3. The unit is not running.

To troubleshoot this problem, proceed as follows:

STEP ONE: Test with a voltmeter at the disconnect box. The voltage reading is 240 volts at L1 and L2, and zero volts at T1 and T2. Proceed to step two.

STEP TWO: You determine that a fuse is blown. Disconnect the power to the unit, remove the wires from the compressor terminals, and test with an ohmmeter between the suction line and the C terminal of the compressor. The meter indicates continuity.

Which component has failed?

UNIT ELEVEN SUMMARY

A common heating/cooling unit found in residential applications is the packaged unit. The heat source for this type of unit can be either a gas heat system, electric heating elements, or a reverse cycle refrigeration system.

With a refrigeration/gas heat combination unit, the heating section is made up of gas burners, a heat exchanger and gas valve, and an ignition system. In both the heating and cooling modes, the air flow pattern is the same. This type of unit is often referred to as a *gas pack*.

When a refrigeration system is used in conjunction with electric heat, the heating elements are located in the path of air flow to the conditioned space.

Heat pumps, in addition to using the reverse refrigeration cycle to accomplish the heating mode, also use resistance heating elements as a supplementary heating source.

Packaged air conditioning systems can be either roof-mounted or pad-mounted.

The split system is another type of system commonly found in residential applications. The condensing unit, located outside, contains the compressor and condenser, while the indoor portion of the system contains the air handling system and the indoor coil. A common indoor coil is the "A" coil.

Gas furnaces are designed as either downflow or upflow types. The upflow gas furnace is popular in standard construction, while the downflow furnace is commonly found in mobile or manufactured homes.

Two methods of ignition used in gas furnaces are the standing pilot system and the electronic ignition system. The standing pilot system uses a thermocouple to keep the gas valve ready to allow gas flow to the main burner, and the electronic ignition system ignites the pilot upon a call for heat from the thermostat. Three types of electronic ignition systems are the glow coil, spark, and hot surface ignitor.

In electric furnaces, up to five resistance heating elements are used to provide heat. A sequencing device is used to bring the elements on one at a time. Electric furnaces can be found in upflow, downflow, or horizontal configurations.

The following pages contain troubleshooting information for various heating, air conditioning, refrigeration, and indoor air quality products. You may wish to use these charts as a basis for developing your own troubleshooting strategy. Remember, always refer to the manufacturer's service literature for specific information regarding the unit you are servicing. Also, use your senses and the various tools and test equipment we have covered to fully evaluate a system before condemning its components. Above all, remember that listening to the customer's complaint is the first step in solving any HVAC-R problem.

Additional troubleshooting charts can be found in Delmar's *Refrigeration and Air Conditioning Technology*, 3rd Edition.

REFRIGERATION TROUBLESHOOTING CHART

PROBLEM	POSSIBLE CAUSES	POSSIBLE CORRECTIVE STEPS
Compressor will not run	1. Main switch open. 2. Fuse blown. 3. Thermal overloads tripped. 4. Defective contactor or coil. 5. System shut down by safety devices. 6. No cooling required. 7. Liquid line solenoid will not open. 8. Motor electrical trouble. 9. Loose wiring.	1. Close switch. 2. Check electrical circuits and motor winding for shorts or grounds. Investigate for possible overloading. Replace fuse after fault is corrected. 3. Overloads are automatically reset. Check unit closely when unit comes back on line. 4. Repair or replace. 5. Determine type and cause of shutdown and correct it before resetting safety switch. 6. None. Wait until calls for cooling. 7. Repair or replace coil. 8. Check motor for open windings, short circuit or burn out. 9. Check all wire junctions. Tighten all terminal screws.
Compressor noisy or vibrating	1. Flooding of refrigerant into crankcase. 2. Improper piping support on suction or liquid line. 3. Worn compressor.	1. Check setting of expansion valves. 2. Relocate, add or remove hangers. 3. Replace.
High discharge pressure	1. Non-condensables in system. 2. System overcharges with refrigerant. 3. Discharge shutoff valve partially closed. 4. Fan not running. 5. Head pressure control setting. 6. Dirty condenser coil.	1. Remove the non-condensables. 2. Remove excess. 3. Open valve. 4. Check electrical circuit. 5. Adjust. 6. Clean.
Low discharge pressure	1. Faulty condenser temperature regulation. 2. Suction shutoff valve partially closed. 3. Insufficient refrigerant in system. 4. Low suction pressure. 5. Variable head pressure valve.	1. Check condenser control operation. 2. Open valve. 3. Check for leaks. Repair and add charge. 4. See corrective steps for low suction pressure. 5. Check valve setting.
High suction pressure	1. Excessive load. 2. Expansion valve overfeeding.	1. Reduce load or add additional equipment. 2. Check remote bulb. Regulate superheat.
Low suction pressure	1. Lack of refrigerant. 2. Evaporator dirty or iced. 3. Clogged liquid line filter drier. 4. Clogged suction line or compressor suction gas strainers. 5. Expansion valve malfunctioning. 6. Condensing temperature too low. 7. Improper TXV.	1. Check for leaks. Repair and add charge. 2. Clean. 3. Replace cartridge(s). 4. Clean strainers. 5. Check and reset for proper superheat. 6. Check means for regulating condensing temperature. 7. Check for proper sizing.
Little or no oil pressure	1. Clogged suction oil strainer. 2. Excessive liquid in crankcase. 3. Low oil pressure safety switch defective. 4. Worn oil pump. 5. Oil pump reversing gear stuck in wrong position. 6. Worn bearings. 7. Low oil level. 8. Loose fitting on oil lines. 9. Pump housing gasket leaks.	1. Clean. 2. Check crankcase heater. Reset expansion valve for higher superheat. Check liquid line solenoid valve operation. 3. Replace. 4. Replace. 5. Reverse direction of compressor rotation. 6. Replace compressor. 7. Add oil and/or through defrost. 8. Check and tighten system. 9. Replace gasket.
Compressor loses oil	1. Lack of refrigerant. 2. Excessive compression ring blowby. 3. Refrigerant flood back. 4. Improper piping or traps.	1. Check for leaks and repair. Add refrigerant. 2. Replace compressor. 3. Maintain proper superheat at compressor. 4. Correct piping.
Compressor thermal protector switch open.	1. Operating beyond design conditions. 2. Discharge valve partially shut. 3. Blown valve plate gasket. 4. Dirty condenser coil. 5. Overcharged system.	1. Add facilities so that conditions are within allowable limits. 2. Open valve. 3. Replace gasket. 4. Clean coil. 5. Reduce charge.

(Courtesy Heatcraft)

REFRIGERATION TROUBLESHOOTING CHART

Maintenance

Evaporators

All evaporator units should be checked once a month or more often for proper defrosting because the amount and pattern of frosting can vary greatly. It is dependent on the temperature of the room, the type of product being stored, how often new product is brought into the room and percentage of time the door to the room is open. It may be necessary to periodically change the number of defrost cycles or adjust the duration of defrost.

Condensing Units / Evaporators

Under normal usage conditions, maintenance should cover the following items at least once every six months.

1. Check and Tighten **ALL** electrical connections.
2. Check all wiring and insulators.
3. Check contactors for proper operation and for worn contact points.
4. Check all fan motors. Tighten motor mount bolts/nuts and tighten fan set screws.
5. Clean the condenser coil surface.
6. Check the refrigerant and oil level in the system.
7. Check the operation of the control system. Make certain all safety controls are operating properly.
8. Check that all defrost controls are functioning properly.
9. Clean the evaporator coil surface.
10. Clean the drain pan and check the drain pan and drain line for proper drainage.
11. Check the drain line heater for proper operation, cuts and abrasions.
12. Check and tighten all flare connections.

Table 18. Evaporator Troubleshooting Chart

SYMPTOMS	POSSIBLE CAUSES	POSSIBLE CORRECTIVE STEPS
Fan(s) will not operate.	1. Main switch open. 2. Blown fuses. 3. Defective motor. 4. Defective Timer or defrost thermostat. 5. Unit in defrost cycle. 6. Coil does not get cold enough to reset thermostat.	1. Close switch. 2. Replace fuses. Check for short circuits or overload conditions. 3. Replace motor. 4. Replace defective component. 5. Wait for completion of cycle. 6. Adjust fan delay setting of thermostat. See Defrost Thermostat Section of this bulletin.
Room temperature too high.	1. Room thermostat set too high. 2. Superheat too high. 3. System low on refrigerant. 4. Coil iced-up.	1. Adjust thermostat. 2. Adjust thermal expansion valve. 3. Add refrigerant. 4. Manually defrost coil. Check defrost controls for malfunction.
Ice accumulating on ceiling around evaporator and/or on fan guards venturi or blades.	1. Defrost duration is too long. 2. Fan delay not delaying fans after defrost period. 3. Defective defrost thermostat or Timer. 4. Too many defrosts.	1. Adjust defrost termination thermostat. 2. Defective defrost thermostat or not adjusted properly. 3. Replace defective component. 4. Reduce number of defrosts.
Coil not clearing of frost during defrost cycle.	1. Coil temperature not getting above freezing point during defrost. 2. Not enough defrost cycles per day. 3. Defrost cycle too short. 4. Defective Timer or defrost thermostat.	1. Check heater operation. 2. Adjust Timer for more defrost cycles. 3. Adjust defrost thermostat or Timer for longer cycle. 4. Replace defective component.
Ice accumulating in drain pan	1. Defective heater. 2. Unit not pitched properly. 3. Drain line plugged. 4. Defective drain line heater. 5. Defective Timer or thermostat.	1. Replace heater. 2. Check and adjust if necessary. 3. Clean drain line. 4. Replace heater. 5. Replace defective component.

(Courtesy Heatcraft)

REMOTE CONDENSING UNIT TROUBLESHOOTING CHART

Complaint	No Cooling							Unsatisfactory Cooling					System Operating Pressures				Test Method Remedy
POSSIBLE CAUSE **DOTS IN ANALYSIS GUIDE INDICATE "POSSIBLE CAUSE"** / **SYMPTOM**	System will not start	Compressor will not start - fan runs	Compressor and Condenser Fan will not star	Evaporator fan will not start	Condenser fan will not start	Compressor runs - goes off on overload	Compressor cycles on overload	System runs continuously - little cooling	Too cool and then too warm	Not cool enough on warm days	Certain areas to cool others to warm	Compressor is noisy	Low suction pressure	Low head pressure	High suction pressure	High head pressure	
Pow er Failure	●																Test Voltage
Blow n Fuse	●		●	●													Impact Fuse Size & Type
Loose Connection	●			●		●											Inspect Connection - Tighten
Shorted or Broken Wires	●	●	●	●	●	●											Test Circuits With Ohmmeter
Open Overload	●	●		●	●												Test Continuity of Overloads
Faulty Thermostat	●			●					●								Test continuity of Thermostat & Wiring
Faulty Transformer	●		●														Check control circuit w ith voltmeter
Shorted or Open Capacitor		●			●	●											Test Capacitor
Internal Overload Open	●																Test Continuity of Overload
Shorted or Grounded Compressor		●				●											Test Motor Windings
Compressor Stuck	●					●											Use Test Cord
Faulty Compressor Contactor	●	●			●	●											Test continuity of Coil & Contacts
Faulty Fan Relay				●													Test continuity of Coil And Contacts
Open Control Circuit				●													Test Control Circuit w ith Voltmeter
Low Voltage		●				●	●										Test Voltage
Faulty Evap. Fan Motor				●									●				Repair or Replace
Shorted or Grounded Fan Motor					●											●	Test Motor Windings
Improper Cooling Anticipator						●	●		●								Check resistance of Anticipator
Shortage of Refrigerant							●	●					●	●			Test For Leaks, Add Refrigerant
Restricted Liquid Line							●	●					●	●			Replace Restricted Part
Undersized Liquid Line								●		●			●				Replace Line
Undersized Suction Line													●				Replace Line
Dirty Air Filter								●		●	●		●				Inspect Filter-Clean or Replace
Dirty Evaporator Coil								●		●	●		●				Inspect Coil - Clean
Not enough air across Evap Coil								●		●	●		●				Speed Blow er, Check Duct Static Press
Too much air across Evap Coil															●		Reduce Blow er Speed
Overcharge of Refrigerant						●	●					●			●	●	Release Part of Charge
Dirty Condenser Coil						●	●			●						●	Inspect Coil - Clean
Noncondensibles							●			●						●	Remove Charge, Evacuate, Recharge
Recirculation of Condensing Air							●			●						●	Remove Obstruction to Air Flow
Infiltration of Outdoor Air								●		●	●						Check Window s, Doors, Vent Fans, Etc.
Improperly Located Thermostat						●			●								Relocate Thermostat
Air Flow Unbalanced									●		●						Readjust Air Volume Dampers
System Undersized								●		●							Refigure Cooling Load
Broken Internal Parts												●					Replace Compressor
Broken Values												●					Test Compressor Efficiency
Inefficient Compressor								●						●	●		Test Compressor Efficiency
High Pressure Control Open			●														Reset And Test Control
Unbalanced Pow er, 3PH		●				●	●										Test Voltage
Wrong Type Expansion Valve						●	●			●							Replace Valve
Expansion Valve Restricted						●	●	●		●			●	●			Replace Valve
Oversized Expansion Valve												●			●		Replace Valve
Undersized Expansion Valve						●	●	●		●			●				Repalce Valve
Expansion Valve Bulb Loose												●			●		Tighten Bulb Bracket
Inoperative Expansion Valve						●		●					●				Check Valve Operation
Loose Hold-dow n Bolts												●					Tighten Bolts

(Courtesy Amana Corporation)

EVAPORATIVE COOLER TROUBLESHOOTING CHART

PROBLEM	POSSIBLE CAUSE	REMEDY
□ **Failure to start or no air delivery**	□ No electrical power to unit: • Fuse blown • Circuit breaker tripped • Electric cord damaged □ Belt too loose or too tight □ Motor overheated and locked: • Belt too tight • Dry bearing on blower wheel	□ Check power: • Replace fuse • Re-set breaker • Replace □ Adjust belt tension □ Replace motor: • Adjust belt tension • Oil blower bearings
□ **Inadequate air delivery with cooler running**	□ Insufficient air exhaust □ Belt too loose □ Pads plugged: • Insufficient water flow over pads	□ Open windows or doors to increase air flow □ Adjust belt tension or replace if needed □ Replace pads: • Clean distributor system
□ **Inadequate cooling**	□ Inadequate exhaust in house □ Pads not wet: • Pads plugged • Dry spots in pads • Dist. tube holes clogged • Pump not working	□ Open windows or doors to increase air flow. □ Check water distribution system: • Clean pads • Clean • Clean • Replace pump (TURN OFF POWER)
□ **Motor cycles on and off**	□ Excessive belt tension □ Blower shaft tight or locked	□ Adjust belt tension □ Oil or replace bearing (TURN OFF POWER)
□ **Noisy**	□ Bearings dry □ Wheel rubbing blower housing	□ Oil bearings □ Inspect and realign (POWER OFF)
□ **Excessive humidity in house**	□ Inadequate exhaust	□ Open doors or windows
□ **Musty or unpleasant odor**	□ Stale or stagnant water in cooler	□ Drain and clean pan

(Courtesy Champion Cooler Corporation)

REFRIGERATOR TROUBLESHOOTING CHART

COMPLAINT	CAUSE - REMEDY
Compressor will not run.	1. No voltage at wall receptacle - check circuit breaker or fuse. 2. Service cord pulled out of wall receptacle - replace. 3. Low voltage causing compressor to cycle on overload. (Voltage fluctuation should not exceed 10% plus or minus from nominal rating.) 4. Control thermostat dial in "Off" position - turn control on. 5. Inoperative control thermostat - replace control. 6. Compressor stuck - replace compressor. 7. Compressor windings open - replace compressor. 8. Defrost timer stuck in defrost - replace defrost timer. 9. Compressor overload stuck open - replace overload. 10. Relay lead loose - repair or replace lead. 11. Relay loose or inoperative - replace relay. 12. Service cord pulled out of harness - repair connection. 13. Faulty cabinet wiring - repair wiring.
Compressor runs but no refrigeration.	1. System out of refrigerant - check for leaks. 2. Compressor not pumping - replace compressor. 3. Restricted filter drier - replace filter drier. 4. Restricted capillary tube - replace. 5. Moisture in system - check for leak in low side.
Compressor short cycles.	1. Erratic control thermostat - replace control. 2. Faulty relay - replace relay. 3. Restricted air flow over condenser - ensure condenser has unobstructed air flow. 4. Low voltage - fluctuation exceeds 10%. (Call qualified electrician.) 5. Compressor draws excessive wattage - replace compressor.
Compressor runs too much or 100%.	1. Erratic control thermostat, or set too cold - replace or reset to normal position. 2. Refrigerator exposed to unusual heat - relocate refrigerator. 3. Abnormally high room temperature - advise customer. 4. Low pumping capacity compressor - replace compressor. 5. Door gaskets not sealing - adjust or replace necessary parts. 6. System undercharged - check for leaks. 7. System overcharged - correct charge. 8. Interior light stays on - check door or lid switch. 9. Non-condensables in system - replace filter drier,evacuate, and recharge. 10. Capillary tube kinked or partially restricted - replace heat exchanger. 11. Filter drier partially restricted - replace filter drier. 12. Excessive service load - advise customer. 13. Restricted air flow over condenser - ensure condenser has unobstructed air flow
Noisy.	1. Tubing vibrates - adjust tubing. 2. Internal compressor noise - replace compressor. 3. Compressor vibrating on cabinet frame - adjust compressor. 4. Loose parts - check shelving, kickplate, defrost drain pan. 5. Compressor operating at high head pressure due to restricted air flow over condenser - ensure condenser has unobstructed air flow.

(Courtesy White Consolidated, Inc.)

REFRIGERATOR TROUBLESHOOTING CHART

COMPLAINT	CAUSE - REMEDY
Freezer compartment too warm.	1. Inoperative fan motor - check wiring and fan motor. 2. Improperly positioned fan - position blade as specified (see Section C.) 3. Evaporator iced up - check defrost system. 4. Defrost heater inoperative - check wiring and defrost heater. 5. Inoperative defrost timer - check wiring and defrost timer. 6. Inoperative defrost thermostat - check wiring and defrost thermostat. 7. Wire loose at defrost timer - repair wire. 8. Excessive service load - advise customer. 9. Abnormally low room temperatures - advise customer. 10. Freezer or refrigerator compartment doors left open - advise customer. 11. Control thermostat out of calibration - replace control. 12. Door gaskets not sealing - adjust or replace necessary parts. 13. Control thermostat sensing element improperly positioned - reposition sensing element. 14. Shortage of refrigerant - check for leaks. 15. Restricted filter drier or capillary tube - check for leaks or "burned" compressor windings.
Refrigerator compartment too warm.	1. Inoperative fan motor - check wiring and fan motor. 2. Improperly positioned fan - position blade as specified (see Section C.) 3. Fan cover missing - replace fan cover. 4. Refrigerator compartment inlet air duct restricted - clean air duct 5. Freezer compartment return air duct restricted - clean air duct. 6. Abnormally low room temperature - advise customer. 7. Control thermostat out of calibration - replace control. 8. Control thermostat knob set at warm setting - set colder. 9. Control thermostat thermal element improperly positioned - reposition sensing element. 10. Evaporator iced up - check defrost system. 11. Inoperative defrost timer - check wiring and defrost timer. 12. Inoperative defrost heater - check wiring and defrost heater. 13. Inoperative defrost thermostat - check wiring and defrost thermostat. 14. Excessive service load - advise customer. 15. Refrigerator compartment or freezer compartment door left open-advise customer. 16. Inoperative or erratic refrigerator compartment and/or freezer compartment door switch - replace door switch. 17. Shortage of refrigerant - check for leaks. 18. Restricted capillary tube or filter drier - check for leaks or "burned" compressor windings.
Evaporator blocked with ice.	1. Inoperative defrost timer - check wiring and defrost timer. 2. Defrost thermostat terminates too early - check for correct positioning of defrost thermostat or replace. 3. Defrost timer incorrectly wired - check wiring. 4. Inoperative fan motor - check wiring and fan motor. 5. Inoperative defrost thermostat - check wiring and defrost thermostat. 6. Inoperative defrost heater - check wiring and defrost heater. 7. Freezer or refrigerator compartment doors left open - advise customer. 8. Freezer defrost drain plugged - clean drain.

(Courtesy White Consolidated, Inc.)

WHITE RODGERS 50E47 HOT SURFACE IGNITION SYSTEM TROUBLESHOOTING CHART

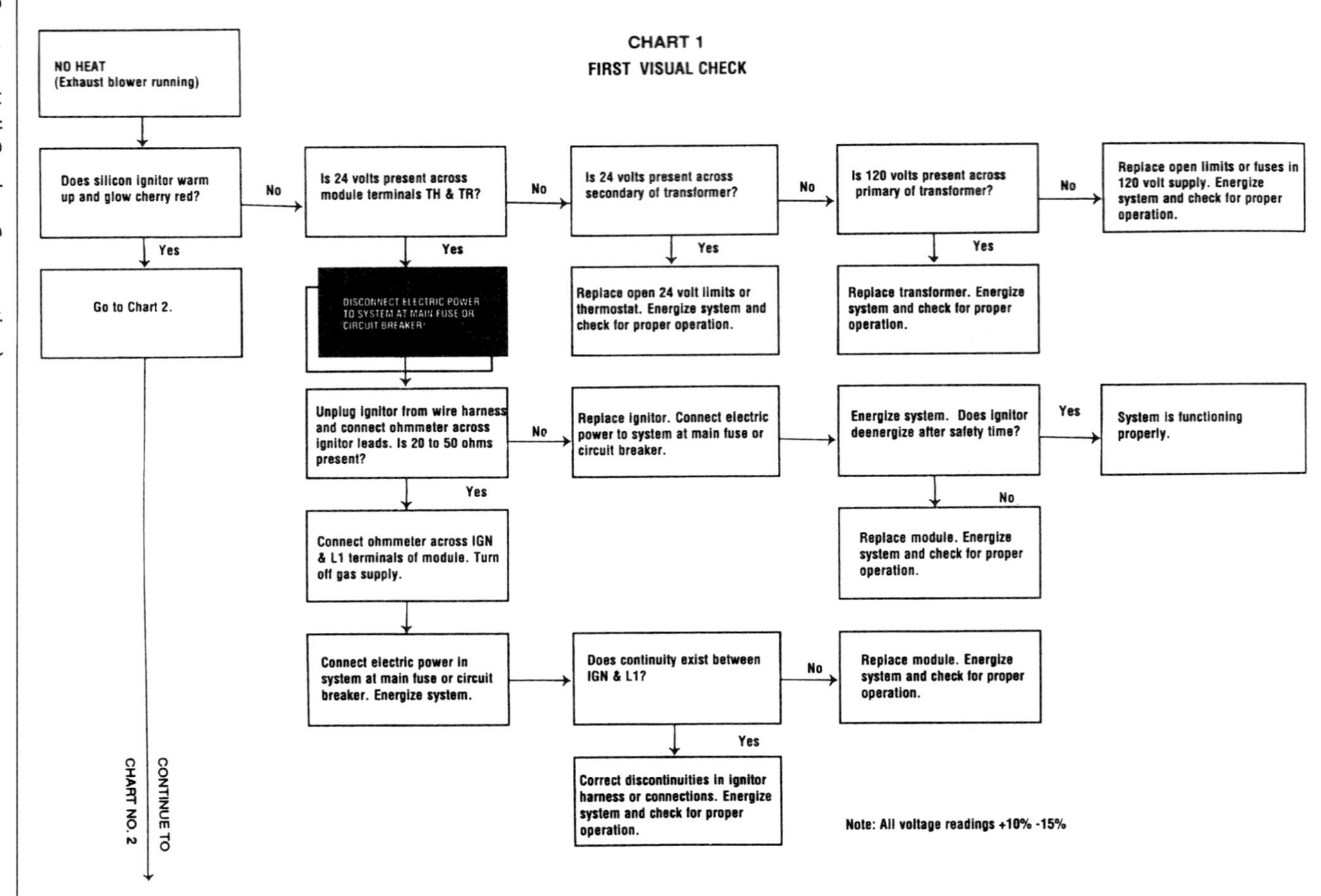

(Courtesy Heil-Quaker Corporation)

WHITE RODGERS 50E47 HOT SURFACE IGNITION SYSTEM TROUBLESHOOTING CHART

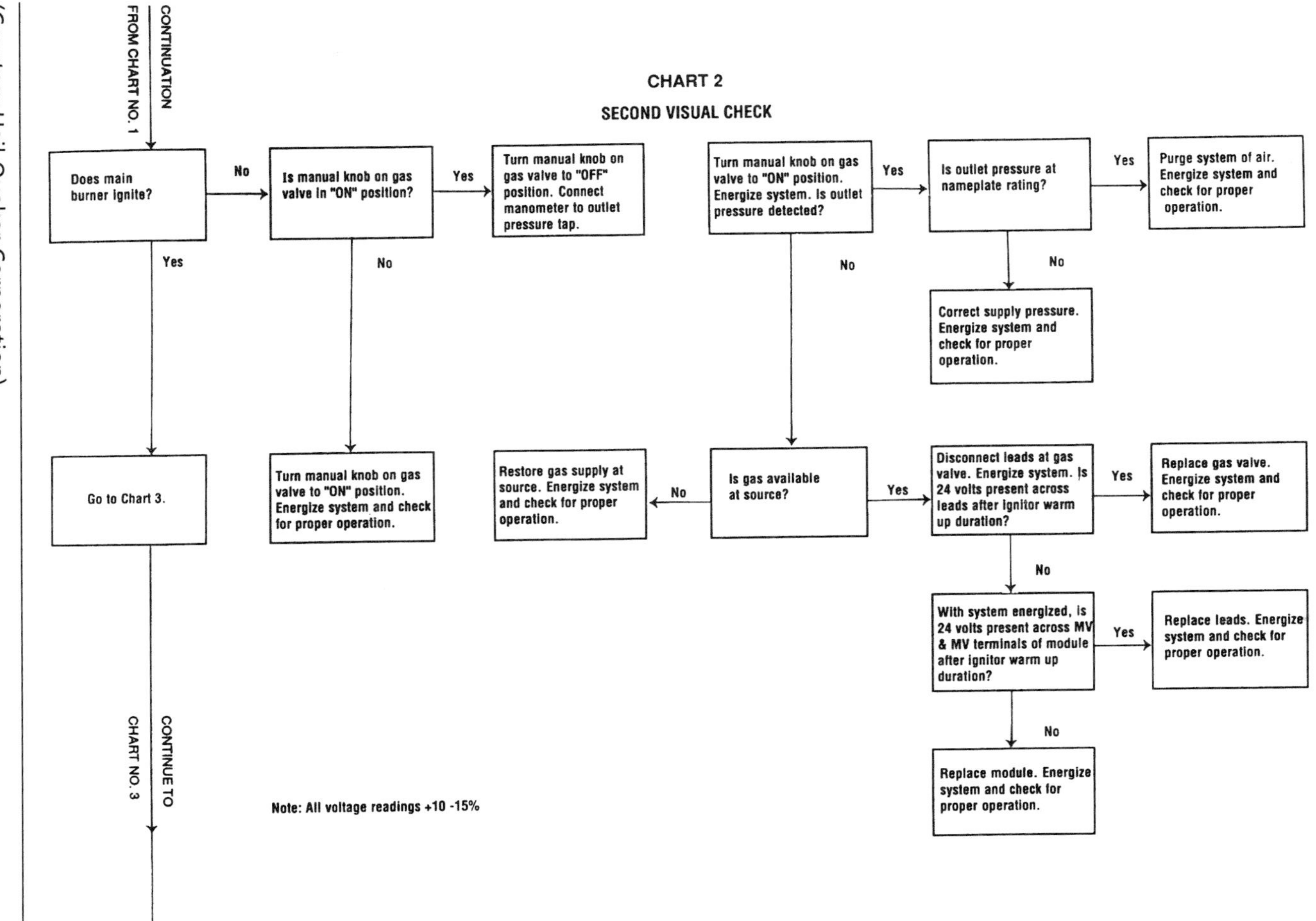

(Courtesy Heil-Quaker Corporation)

WHITE RODGERS 50E47 HOT SURFACE IGNITION SYSTEM TROUBLESHOOTING CHART

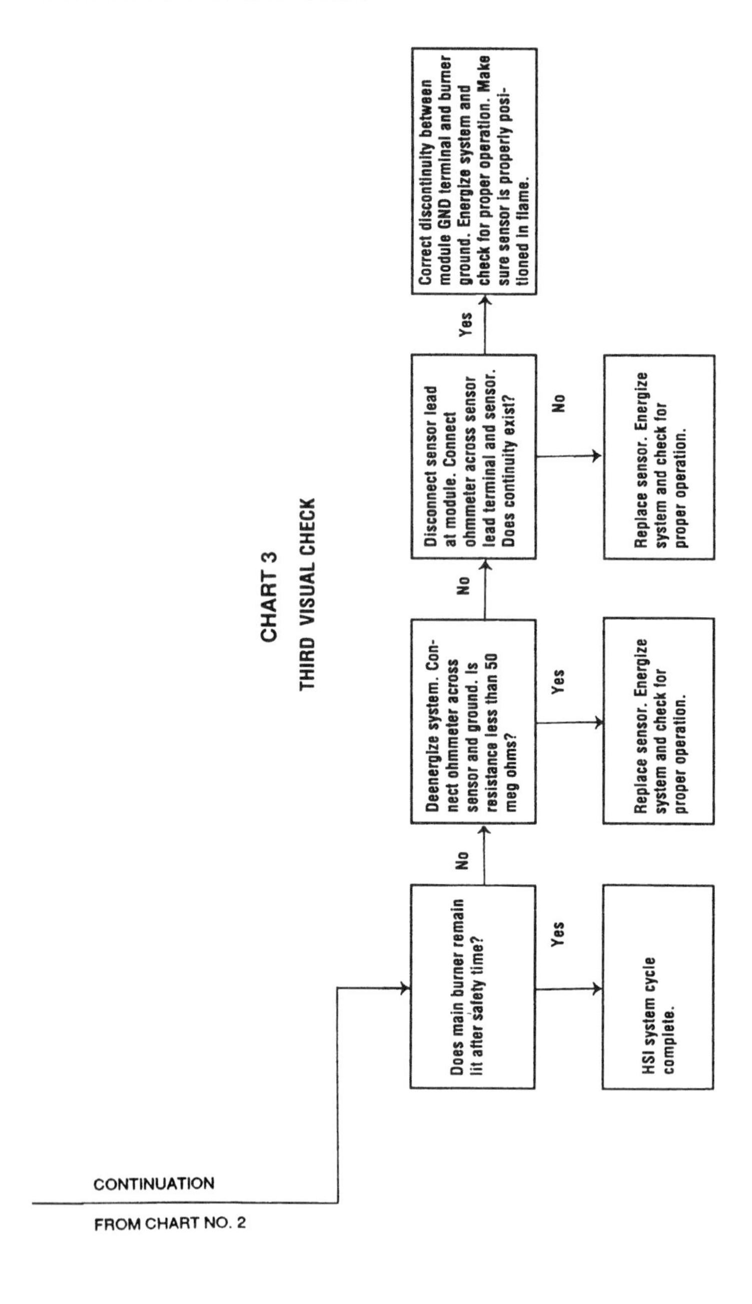

(Courtesy Heil-Quaker Corporation)

HONEYWELL S86 SPARK-TO-PILOT IGNITION SYSTEM TROUBLESHOOTING CHART

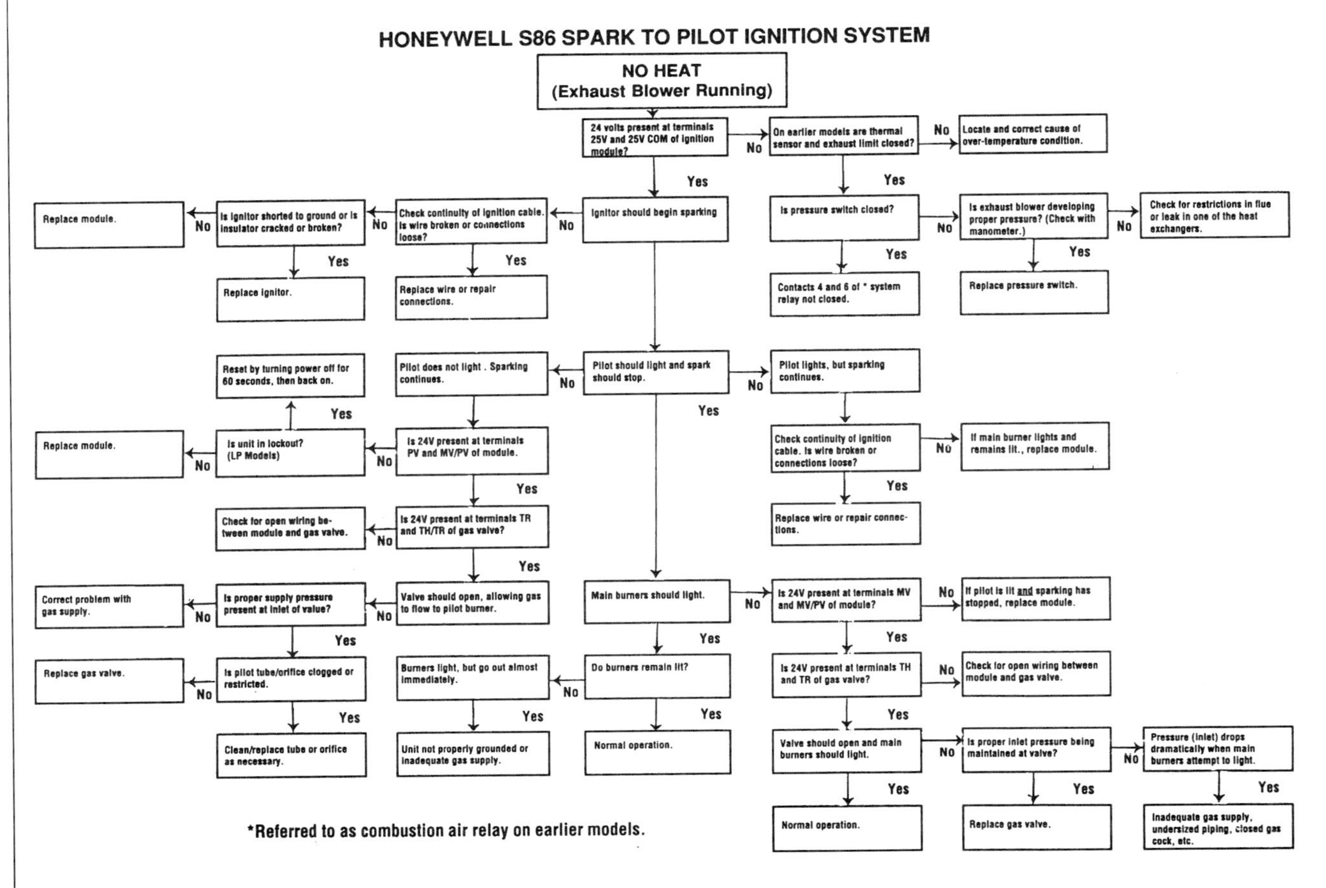

(Courtesy Heil-Quaker Corporation)

CONDENSING GAS FURNACE TROUBLESHOOTING CHART

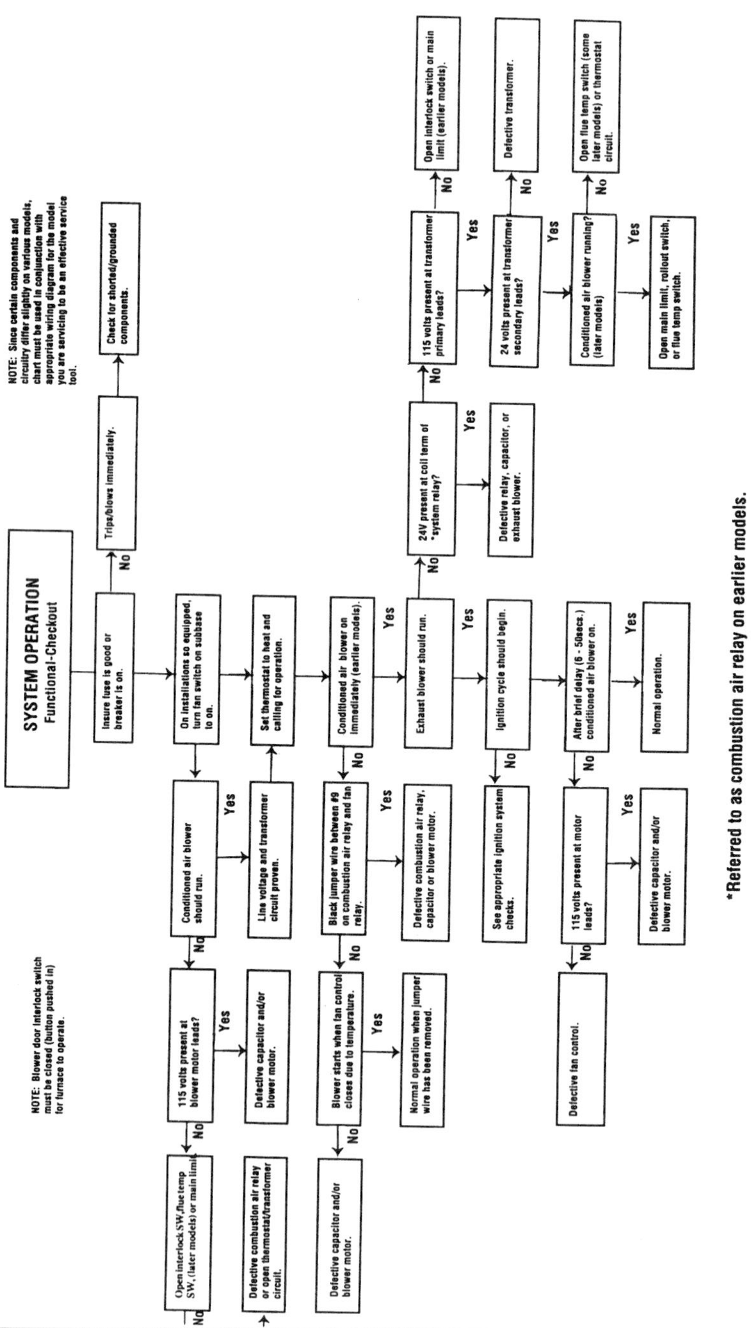

(Courtesy FHP)

DIRECT GAS-FIRED INDUSTRIAL AIR HEATER TROUBLESHOOTING CHART

PROBLEM	POSSIBLE CAUSE	CORRECTIVE ACTION
I. NO BLOWER OPERATION	1. Mode Selector Switch a) Switch in OFF position	a) Place switch in proper mode
	2. Blower Service Switch a) Switch in OFF position b) Defective switch	a) Place switch in REMOTE position b) Replace switch
	3. Control Transformer a) No input voltage b) Blown control fuse c) Defective transformer	a) Check disconnect and supply fusing b) Replace control fuse c) Replace transformer
	4. Unit in Reset a) Time Delay relay timed out (Outside temperature below 45°F in Vent Mode or R1 relay not energized on call for heat)	a) Turn unit OFF momentarily and turn unit ON
	5. Damper End Switch a) Switch not made b) Damper not operating c) Defective Damper Motor	a) Adjust end switch setting b) Check damper operation c) Replace damper motor
	6. Motor Protection a) Overload relay tripped b) Overload relay defective	a) Reset overload relay and check motor amps/overload setting b) Replace overload relay
	7. Motor Starter a) Defective starter	a) Replace starter
	8. Motor a) No input voltage b) Improper wiring c) Defective motor	a) Check fusing b) Correct wiring c) Replace Motor
	9. Blower Damage a) Defective or locked bearings. b) Check for physical damage	a) Replace bearings b) Replace or repair blower
	10. Belts a) Belt Slipping b) Belt broken or missing	a) Tighten belts b) Replace belts
	11. Control Relays a) Improper part b) Improper wiring c) Defective relay (CR-1)	a) Check relay voltage (24 volt) b) Check wiring c) Replace relay
	12. Operating Thermostat a) Thermostat satisfied b) Open in thermistor circuit c) Defective thermostat	a) Adjust thermostat, if applicable b) Check wiring or replace thermistor c) Replace thermostat

(Courtesy Cambridge Engineering Inc.)

DIRECT GAS-FIRED INDUSTRIAL AIR HEATER TROUBLESHOOTING CHART

PROBLEM	POSSIBLE CAUSE	CORRECTIVE ACTION
II. BLOWER RUNS; NO HEAT; FLAME SAFETY RELAY DOESN'T LOCK OUT	1. Mode Selector Switch a) Switch in VENT position	a) Place switch in proper mode
	2. Burner Service Switch a) Switch in OFF position	a) Place switch in REMOTE position
	3. Entering Air Thermostat a) Thermostat satisfied	a) Adjust thermostat, if applicable
	4. Airflow Switch a) Blower running backwards b) Belts slipping c) Blocked intake or discharge d) Clogged airflow tubing or pick-up ports e) Switch defective	a) Reverse motor direction b) Tighten and/or replace belts c) Find and remove obstruction d) Clean or replace tubing or pick-up ports e) Replace switch
	5. Flame Safeguard Relay (FSR) a) No input voltage b) Defective FSR	a) Check wiring b) Replace FSR
III. BLOWER RUNS; NO HEAT; FLAME SAFETY RELAY LOCKS OUT	1. Igniter a) No current (open igniter) b) No voltage	During trial for ignition: a) Check igniter current b) Check FSR output to igniter
	2. High Limit a) High limit tripped	a) See Problem Number VI. (P.75)
	3. Gas Valve a) No input voltage b) Gas valve does not open c) Defective solenoid	a.1) Check FSR output to R1 relay during ignition trial a.2) Check gas valve circuit and wiring b.1) Compare supply voltage to nameplate voltage b.2) Inlet gas pressure too high. b.3) Clean and/or replace gas valve parts c) Replace solenoid or valve assembly
	4. Modulating Valve a) Low fire set too low	a) Adjust low fire on modulating valve
	5. Regulator a) Clogged vent orifice b) No supply pressure c) Improper manifold pressure d) Defective regulator	a) Clean or replace orifice b) Check all gas cocks and piping c) Adjust regulator d) Replace regulator
	6. High or Low Gas Pressure Switch a) Gas pressure switch tripped b) Pressure switch defective	a.1) Check gas supply for low gas pressure or no gas a.2) Check manifold gas pressure for high pressure reading b) Replace gas pressure switch

(Courtesy Cambridge Engineering Inc.)

DIRECT GAS-FIRED INDUSTRIAL AIR HEATER TROUBLESHOOTING CHART

PROBLEM	POSSIBLE CAUSE	CORRECTIVE ACTION
IV. BLOWER RUNS; UNIT HEATS; FLAME SAFETY RELAY LOCKS OUT	1. Low Flame Current	
	a) Flame rod oxidized	a) Scrape oxide coating off rod or replace flame rod
	b) Dirt build-up on insulator	b) Clean dirt deposit from insulator surface and install protective boot
	c) Low fire set too low	c) Adjust low fire on modulating valve
	2. No flame Current	
	a) Flame rod oxidized or grounded	a) Replace flame rod
	b) Ground connection open	b.1) Reference transformer to ground b.2) Secure FSR grounded b.3) Tighten loose ground screws
	c) Wire termination oxidized	c) Clean terminal and reinsert
	3. Fluctuating Flame Current	
	a) Unit overfiring	a) Check manifold pressure
	b) Defective Burner	b) Replace burner
	c) Intermittent ground connection	c) Tighten all ground points
	4. R1 Relay	
	a) Welded contact in LTC circuit	a) Replace relay
	5. Flame Safeguard Relay	
	a) Defective FSR	a) Replace relay
V. BLOWER RUNS; UNIT HEATS; SHORT CYCLES WITHOUT RESETTING	1. Air Flow Switch	
	a) Blower running backwards	a) Reverse motor direction
	b) Belts slipping	b) Tighten and/or replace belts
	c) Blocked intake or discharge	c) Find and remove obstruction
	d) Air delivery below unit specs	d) Increase fan RPM for air delivery to comply with minimum requirements
	e) Clogged airflow tubing or pick-up ports	e) Clean or replace airflow tubing or pick-up ports
	f) Defective switch	f) Replace switch
	2. Flame Safeguard Relay	
	a) Defective FSR	a) Replace FSR
	3. Operating Thermostat	
	a) Differential temperature setting too tight	a) Increase differential temperature setting
	4. Damper Motor End Switch	
	a) End switch not adjusted properly	a) Adjust end switch
	5) Entering Air Thermostat	
	a) Differential temperature setting too tight	a) Increase differential temperature setting

(Courtesy Cambridge Engineering Inc.)

DIRECT GAS-FIRED INDUSTRIAL AIR HEATER TROUBLESHOOTING CHART

PROBLEM	POSSIBLE CAUSE	CORRECTIVE ACTION
VI. HIGH LIMIT TRIPPED	1. High Limit	
	a) TDM reading for high limit temperature above 160°F	a) Perform high limit calibration (See Page 34)
	b) High limit will not reset	b) Replace high limit
	2. Unit Overfiring	
	a) TDM reading for discharge temperature with and without burner operating exceeds allowable temperature rise for heater	a) Adjust appliance regulator to obtain temperature rise specified on heater nameplate
	3. Discharge Damper	
	a) Damper blades partially closed	a) Adjust damper linkage
	b) Damper motor defective	b) Replace damper motor
	4. Airflow Restricted	
	a) Blower running backwards	a) Reverse motor direction
	b) Belts slipping	b) Tighten and/or Replace belts
	c) Blocked intake or discharge	c) Find and remove obstruction
	5. Temperature Control System	
	a) Temperature control system does not modulate	a) See Problem VIII for MD systems (below) or Problem IX for ED, EDR, or EDSM (P.76)
	b) System modulates but TDM reading for discharge temperature is above 160°F for prolonged time	b) Perform temperature control system calibration (See Page 34)
VII. BLOWER RUNS; UNIT HEATS; WILL NOT CYCLE OFF	1. Operating Thermostat	
	a) Open in thermistor circuit	a) Check thermistor wiring and/or replace thermistor
	b) Thermostat defective	b) Replace thermostat
	c) Thermostat located improperly	c.1) Thermostat in cold draft-relocate c.2) Thermostat not satisfied-turn down
	d) Thermostat differential setting too wide	d) Reduce differential setting
	2. Burner Service Switch	
	a) Switch in LOCAL position	a) Place switch in REMOTE position
	3. Auxiliary Control	
	a) Auxiliary contacts closed	a) Check auxiliary circuit wiring and contacts
VIII. MODULATING VALVE DOES NOT MODULATE (MD VALVE)	1. Modulating Valve	
	a) Sensing bulb not seated in bulb clips	a) Reseat bulb in mounting clips
	b) Top of valve housing not secured to valve	b) Tighten the two Phillips head screws
	c) Capillary kinked or out of charge	c) Replace valve
	d) Valve out of calibration range	d) Perform temperature control system calibration

(Courtesy Cambridge Engineering Inc.)

DIRECT GAS-FIRED INDUSTRIAL AIR HEATER TROUBLESHOOTING CHART

PROBLEM	POSSIBLE CAUSE	CORRECTIVE ACTION
IX. MODULATING VALVE DOES NOT MODULATE; CONTINUOUS HIGH FIRE (ED, EDR,& EDSM)	1. Amplifier (A1014 or A1044)	
	a) Wire not connected to amplifier terminal 3 or 4. Also terminal 5 on A1044 only.	a) Re-install wire
	b) Jumper not installed between terminals 2 and 3 of A1014 amplifier only.	b) Re-install jumper
	c) Amplifier Defective	c) Replace Amplifier
	2. Discharge Temperature Sensor (TS114 or TS144)	
	a) Open in sensor circuit	a) Replace the sensor if the resistance measured at: terminals 1 and 2 on TS114 sensor exceeds 11,000 Ω; terminals 1 and 3 or 2 and 3 on TS144 exceeds 6,000 Ω.
	b) Temperature control system out of calibration range	b) Perform temperature control system calibration
	c) Sensor cross-wired to amplifier	c) Correct wiring terminations
	3. Space Temperature Selector (T244 or TS244/TD244)	
	a) Open in sensor circuit	a) Replace the sensor if the resistance measured is more than: 7,000 Ω for the T244; 5,500 Ω for the TS244; or 2,250 Ω for the TD244.
	b) Induced voltage in field wiring	b) Utilize shielded, twisted pair wiring. (See Page 21 and 22)
	c) Space sensor in cold draft	c) Relocate sensor
	4. Remote Heat Adjust (RHA) (TD114 or CEI 4175-0-960)	
	a) Short in RHA circuit	a) Replace RHA if resistance measured between terminals 1 and 3 of RHA is less than 6,000 Ω.
	b) Induced voltage in field wiring	b) Utilize shielded, twisted pair wiring (See Page 21 and 22)
	5. Modulating Valve (M611 or MR212)	
	a) Foreign material holding valve open	a) Disassemble valve and remove foreign material.
X. MODULATING VALVE DOES NOT MODULATE; CONTINUOUS LOW FIRE (ED, EDR, & EDSM)	1. Transformer Class II	
	a) No voltage output to amplifier	a) Replace transformer (Also check for short in modulating valve coil)
	2. Modulating Valve	
	a) Valve coil is open or shorted	a) Replace valve coil if its resistance is less than 40 Ω or greater than 85 Ω.
	b) Plunger jammed	b) Clean or replace plunger
	c) Ruptured main or balancing diaphragm	c) Determine diaphragm condition and replace if defective.

(Courtesy Cambridge Engineering Inc.)

DIRECT GAS-FIRED INDUSTRIAL AIR HEATER TROUBLESHOOTING CHART

PROBLEM	POSSIBLE CAUSE	CORRECTIVE ACTION
X. MODULATING VALVE DOES NOT MODULATE; CONTINUOUS LOW FIRE (ED, EDR, & EDSM) Continued:	3. Amplifier a) No output voltage to valve	a) With the wire removed from terminal 3 of amplifier, replace amplifier if the valve voltage does not exceed 18 volts DC
	4. Discharge Temperature Sensor (TS114 or TS144) a) Short in sensor circuit b) Temperature control system out of calibration range	a) Replace the sensor if the resistance measured at: terminals 1 and 2 on TS114 is less than 8000 Ω; terminals 1 and 3 or 2 and 3 on TS144 is less than 2900 Ω. b) Perform temperature control system calibration
	5. Space Temperature Selector (T244 or TS244/TD244) a) Short in sensor circuit	a) Replace the sensor if the resistance measured is less than 5,000 Ω for the T244 or 3,500 Ω for the TS244 and 1,950 Ω for the TD244.
	6. Remote Heat Adjust (RHA) (TD114 or CEI 4175-0-960) a) Open in the RHA control circuit	a) Replace the control if the resistance measured at terminals 1 and 3 exceed 12,000 Ω.
XI. ERRATIC or PULSATING FLAME	1. High Pressure Regulator a) Vent undersized b) Defective regulator	a) Enlarge vent piping size or reduce vent piping length. b) Replace regulator
	2. Amplifier a) Hunting b) Temperature control system out of calibration c) Defective Amplifier	a) Adjust sensitivity control dial counter-clockwise b) Perform temperature control system calibration c) Replace amplifier
	3. Space Temperature Selector (T244 or TS244/TD244) a) Induced voltage in field wiring	a) Utilize shielded, twisted pair wiring. (See Page 21 and 22)
	4. Remote Heat Adjust (RHA) (TD114 or CEI 4175-0-960) a) Induced voltage in field wiring	a) Utilize shielded, twisted pair wiring. (See Page 21 and 22)

(Courtesy Cambridge Engineering Inc.)

WATER-TO-AIR HEAT PUMP TROUBLESHOOTING CHART

PROBLEM	POSSIBLE CAUSE	CHECKS AND CORRECTIONS
ENTIRE UNIT DOES NOT RUN	Blown fuse Broken or loose wires Voltage supply low Thermostat	Replace fuse or reset circuit breaker (Check for correct fuse) Replace or tighten the wires If voltage is below minimum voltage specified on dataplate, contact local power company. Check 24 volt transformer for burnout or voltage less than 18 volts Set thermostat to "COOL" and lowest temperature setting, unit should run. Set thermostat to "HEAT" and highest temperature, unit should run. Set fan to "ON", fan should run. If unit does not run in all 3 cases, the thermostat could be wired incorrectly or faulty. To ensure faulty or miswired thermostat, disconnect thermostat wires at unit and jumper between "R", "Y", "G" and "W" terminals and unit should run in cooling.
BLOWER OPERATES BUT COMPRESSOR DOES NOT	Voltage supply low Thermostat Wiring Safety controls	If voltage is below minimum voltage specified on the dataplate, contact local power company. Check setting, calibration and wiring Check for loose or broken wires at compressor, capacitor or contactor. The unit could be off on the cutout control safety circuit. Reset the thermostat to "OFF". After a few minutes turn to "COOL" or "HEAT". If the compressor runs, unit was in safety control lock out(See problems for possible causes)
UNIT OFF ON HIGH PRESSURE CONTROL	Discharge pressure too high	In "COOLING" mode: Lack of adequate water flow. Entering water too warm. Scaled or plugged condenser. In "HEATING" mode: Lack of adequate air flow. Entering air too hot Blower inoperative, clogged coil or dirty filter, restrictions in ductwork
	Refrigerant charge	The unit is overcharged with refrigerant. Reclaim, evacuate and recharge with specified amount of R-22.
	High pressure switch	Check for defective or improperly calibrated high pressure switch
UNIT OFF ON LOW PRESSURE CONTROL	Suction pressure too low	In "COOLING" mode: Lack of inadequate air flow. Entering air too cold. Blower inoperative, clogged coil or dirty filter, restrictions in duct work. In "HEATING" mode: Lack of adequate water flow. Entering water too cold. Scaled or plugged condenser.
	Refrigerant charge	The unit is low in charge of refrigerant. Locate the leak repair evacuate and recharge with specified amount of R-22
	Low pressure switch	Check for defective or improperly calibrated low pressure switch.
UNIT SHORT CYCLES	Thermostat Wiring and controls Compressor overload	The differential is set too close in the thermostat. Readjust heat anticipator. Loose connections in the wiring or the control contactors defective Defective compressor overload, check and replace if necessary. If the compressor runs too hot it may be due to the deficient refrigerant charge
INSUFFICIENT COOLING OR HEATING	Unit undersized	Recalculate heat gains or losses for space to be conditioned. If excessive rectify by adding insulation shading etc.
	Loss of conditioned air by leaks	Check for leaks in ductwork or introduction of ambient air thru doors and windows
	Thermostat	Improperly located thermostat (e.g. near kitchen sensing inaccuarately the comfort level in living areas)
	Airflow	Lack of adequate airflow or improper distribution of air.
	Refrigerant charge	Low on refrigerant charge causing inefficient operation.
	Compressor	Check for defective compressor, If discharge pressure is too low and suction pressure is too high, compressor is not pumping properly. Replace compressor.
	Reversing valve	Defective reversing valve creating bypass of refrigerant from discharge to suction side of compressor.
	Operating pressure	Incorrect operating pressure (See chart)
	Refrigerant system	Check expansion valve for possible restrictions to flow of refrigerant. The refrigerant system may be contaminated with moisture, noncondensables, and particles. Dehydrate, evacuate and recharge the system

(Courtesy FHP)

HEAT PUMP TROUBLESHOOTING CHART

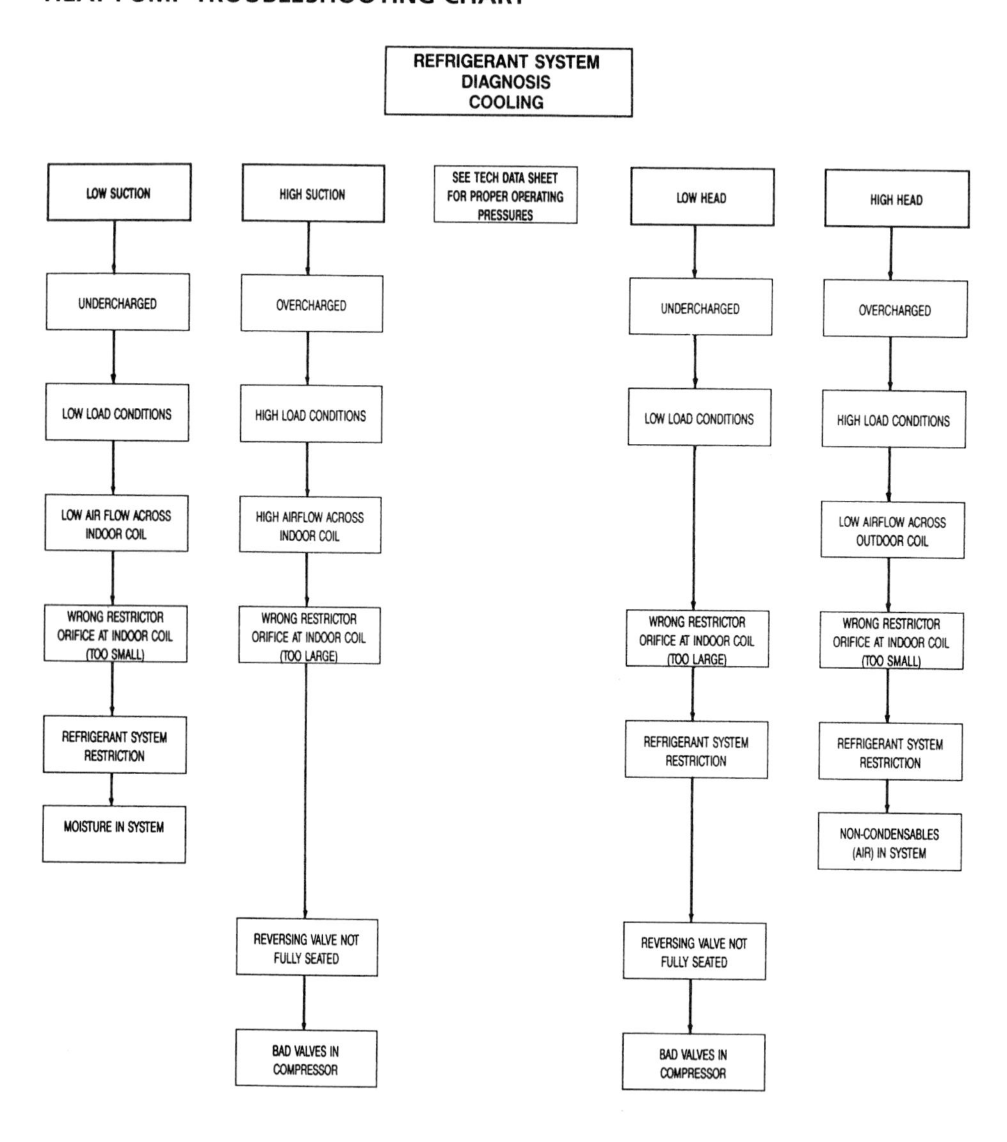

(Courtesy Heil-Quaker Corporation)

HEAT PUMP COOLING/DEFROST CYCLE

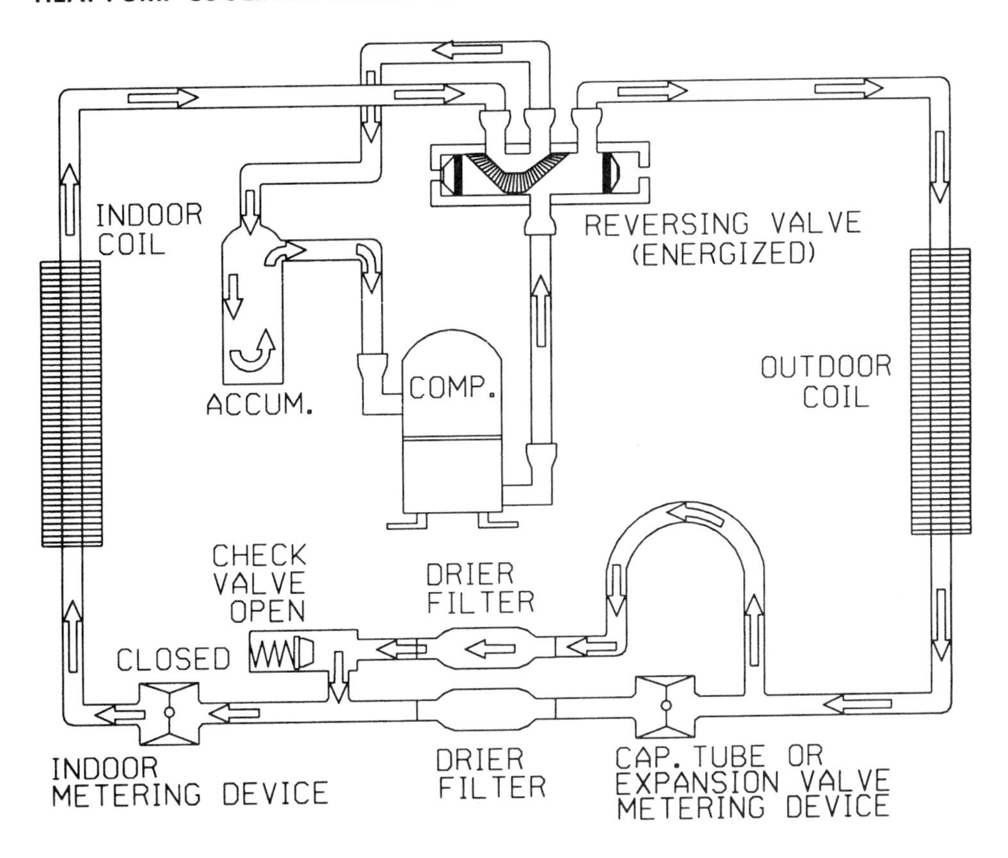

HEAT PUMP TROUBLESHOOTING CHART

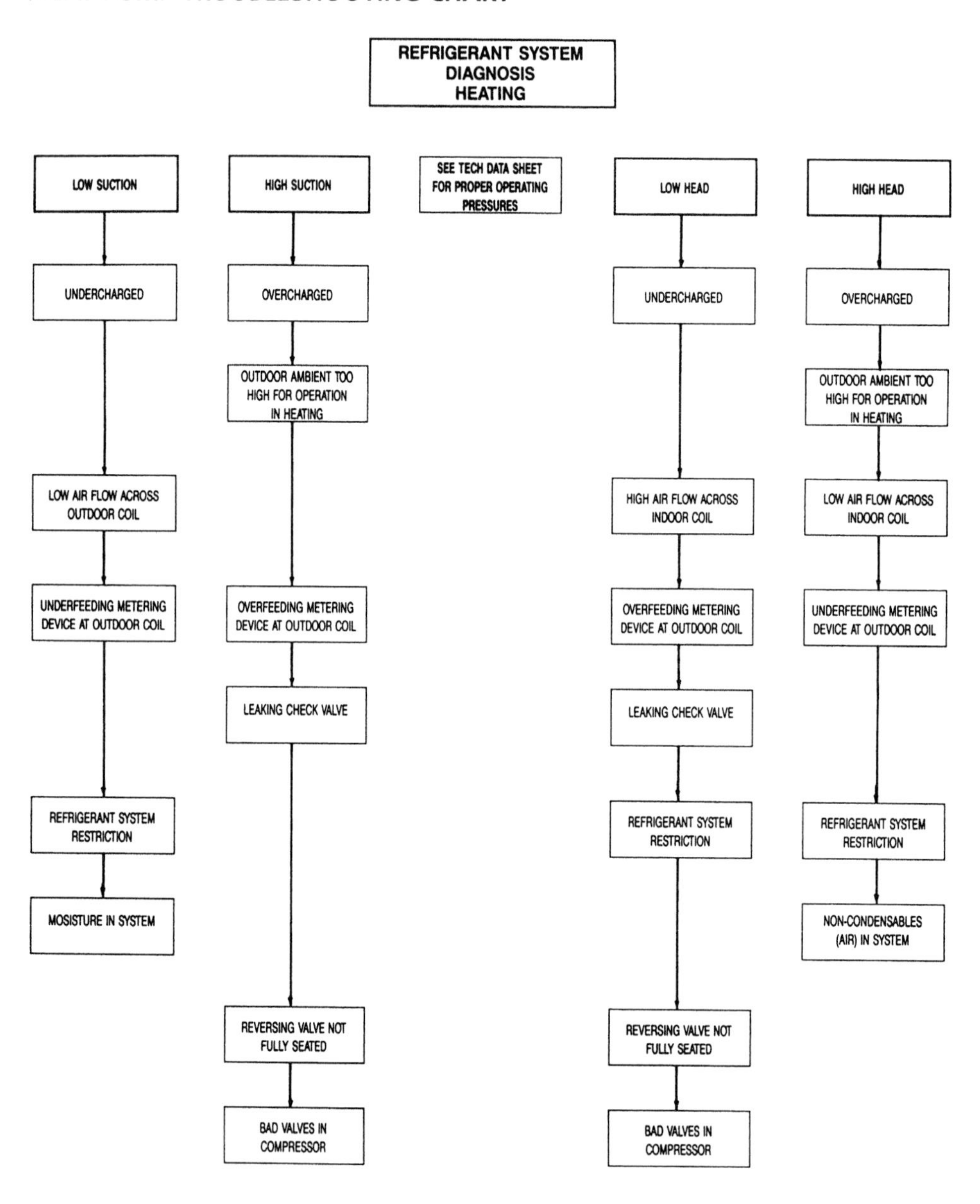

(Courtesy Heil-Quaker Corporation)

HEAT PUMP HEATING CYCLE

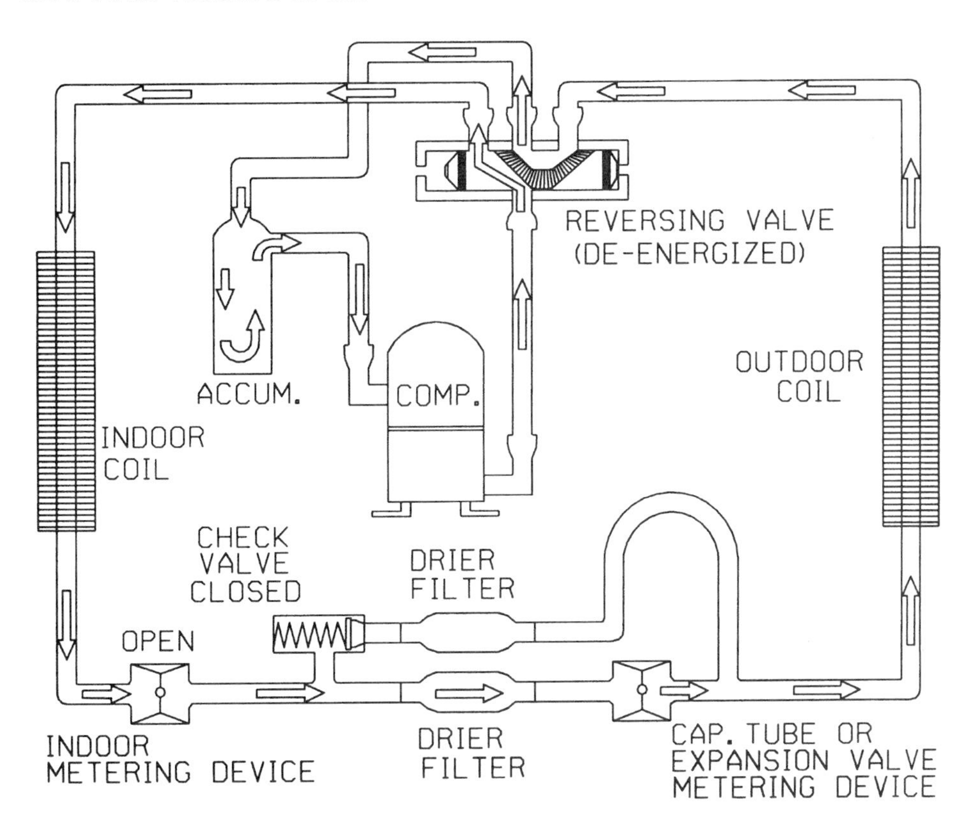

HEAT PUMP TROUBLESHOOTING CHART

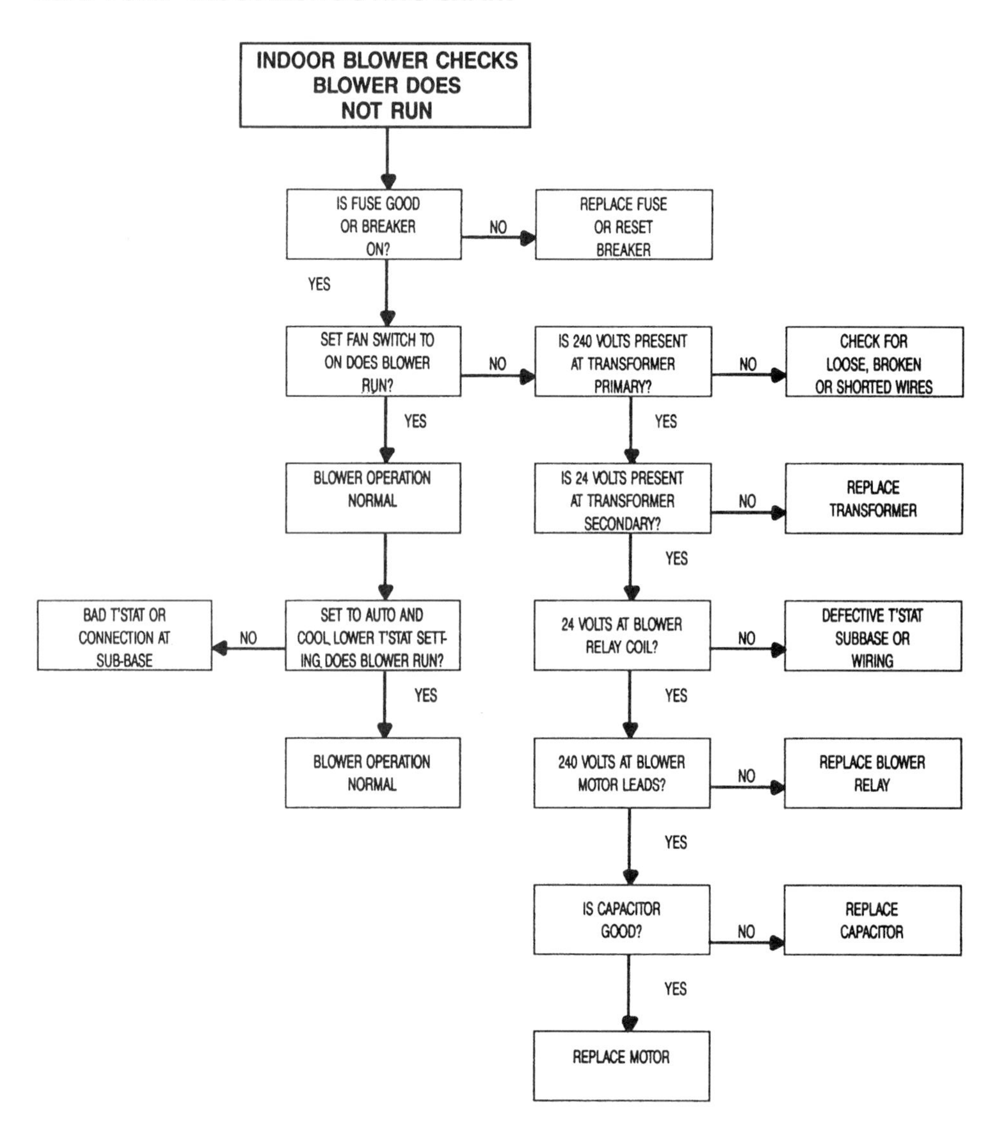

(Courtesy Heil-Quaker Corporation)

HEAT PUMP TROUBLESHOOTING CHART

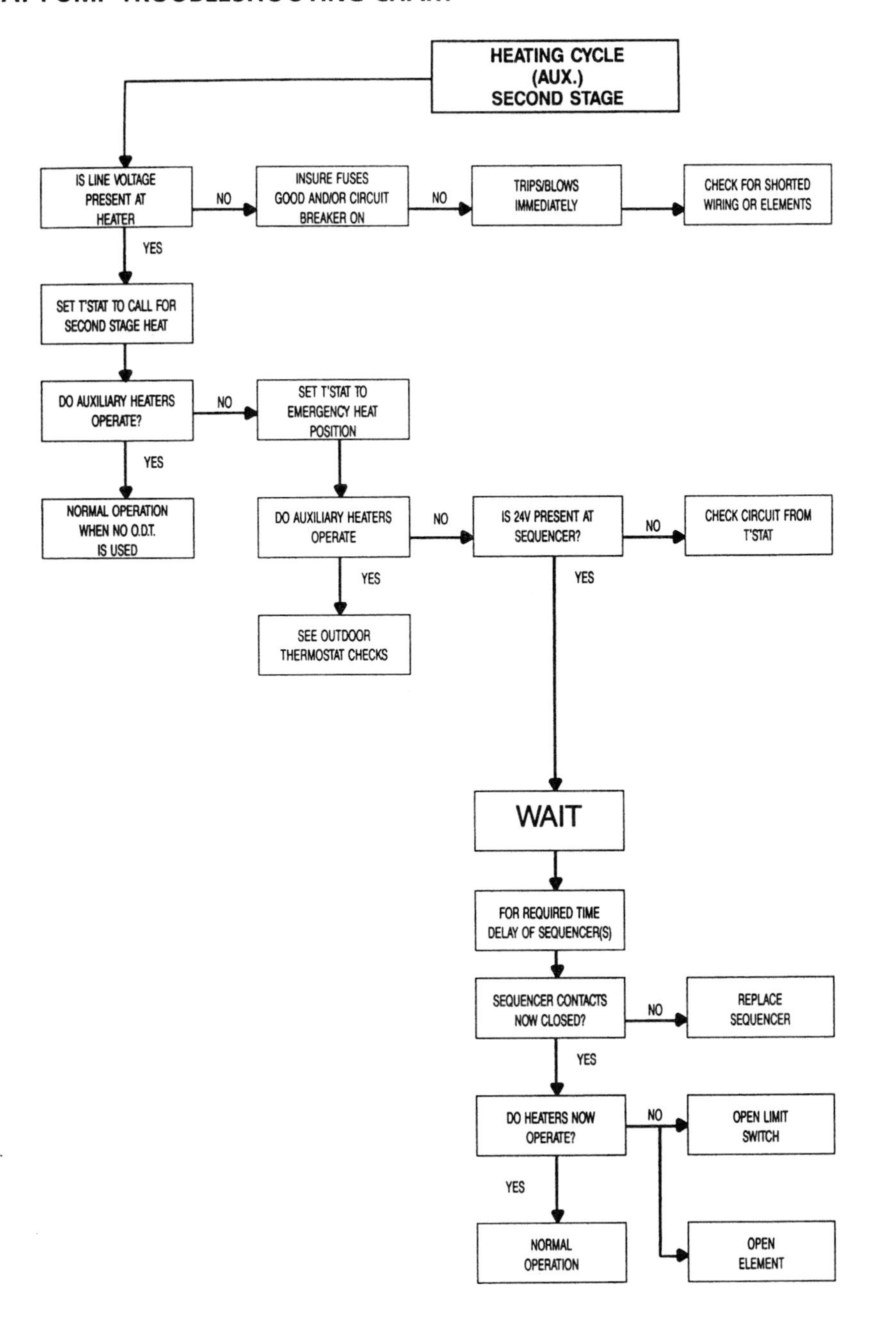

(Courtesy Heil-Quaker Corporation)

HEAT PUMP TROUBLESHOOTING CHART

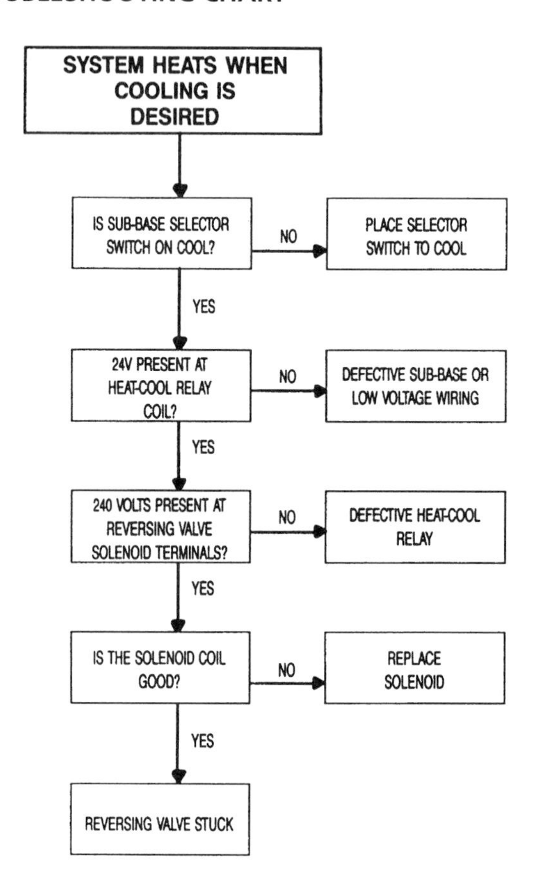

(Courtesy Heil-Quaker Corporation)

HEAT PUMP TROUBLESHOOTING CHART

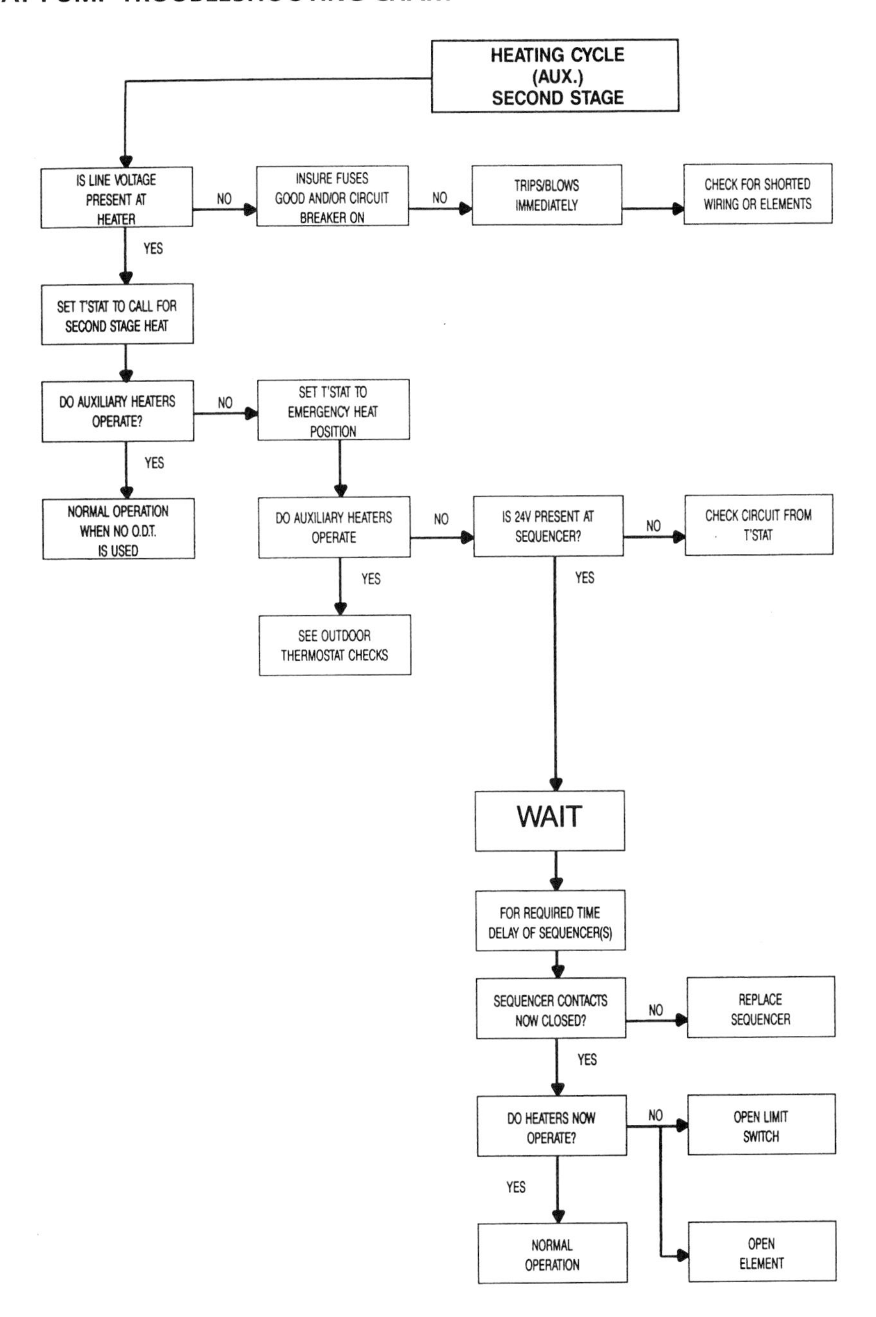

(Courtesy Heil-Quaker Corporation)

HEAT PUMP TROUBLESHOOTING CHART

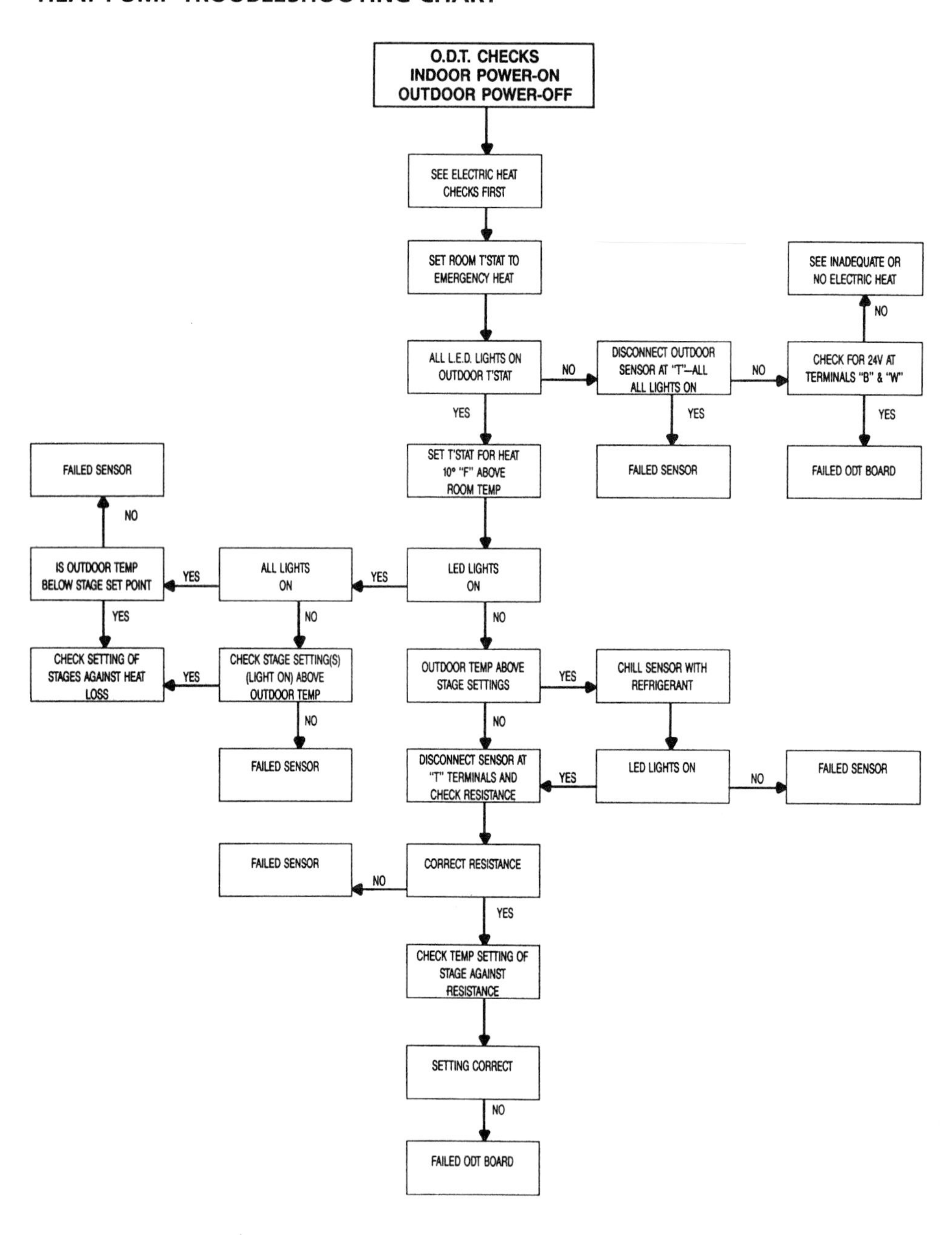

(Courtesy Heil-Quaker Corporation)

REMOTE HEAT PUMP TROUBLESHOOTING CHART

Complaint	No Cooling							Unsatisfactory Cooling					System Operating Pressures				
POSSIBLE CAUSE DOTS IN ANALYSIS GUIDE INDICATE "POSSIBLE CAUSE" — SYMPTOM	System will not start	Compressor will not start - fan runs	Compressor and Condenser Fan will not start	Evaporator fan will not start	Condenser fan will not start	Compressor runs - goes off on overload	Compressor cycles on overload	System runs continuously - little cooling	Too cool and then too warm	Not cool enough on warm days	Certain areas to cool others to warm	Compressor is noisy	Low suction pressure	Low head pressure	High suction pressure	High head pressure	Test Method Remedy
Pow er Failure	●																Test Voltage
Blow n Fuse	●		●	●													Impact Fuse Size & Type
Loose Connection	●			●		●											Inspect Connection - Tighten
Shorted or Broken Wires	●	●	●	●	●	●											Test Circuits With Ohmmeter
Open Overload	●	●		●	●												Test Continuity of Overloads
Faulty Thermostat	●			●					●								Test continuity of Thermostat & Wiring
Faulty Transformer	●		●														Check control circuit w ith voltmeter
Shorted or Open Capacitor		●			●	●											Test Capacitor
Internal Overload Open	●																Test Continuity of Overload
Shorted or Grounded Compressor		●				●											Test Motor Windings
Compressor Stuck	●					●											Use Test Cord
Faulty Compressor Contactor	●	●			●	●											Test continuity of Coil & Contacts
Faulty Fan Relay				●													Test continuity of Coil And Contacts
Open Control Circuit				●													Test Control Circuit w ith Voltmeter
Low Voltage		●				●	●										Test Voltage
Faulty Evap. Fan Motor				●									●				Repair or Replace
Shorted or Grounded Fan Motor					●											●	Test Motor Windings
Improper Cooling Anticipator						●	●		●								Check resistance of Anticipator
Shortage or Refrigerant							●	●					●	●			Test For Leaks, Add Refrigerant
Restricted Liquid Line							●	●					●	●			Replace Restricted Part
Undersized Liquid Line								●		●			●				Replace Line
Undersized Suction Line													●			◆	Replace Line
Dirty Air Filter								●		●	●		●			◆	Inspect Filter-Clean or Replace
Dirty Indoor Coil								●		●	●		●			◆	Inspect Coil - Clean
Not enough air across Indoor Coil								●		●	●		●			◆	Speed Blow er, Check Dust Static Pressu
Too much air across Indoor Coil															●		Reduce Blow er Speed
Overcharge of Refrigerant						●	●					●			●	●	Release Part of Charge
Dirty Outdoor Coil						●	●			●			◆			●	Inspect Coil - Clean
Noncondensibles							●			●						●	Remove Charge, Evacuate, Recharge
Recirculation of Condensing Air							●			●						●	Remove Obstruction to Air Flow
Infiltration of Outdoor Air								●		●	●						Check Window s, Doors, Vent Fans, Etc.
Improperly Located Thermostat						●			●								Relocate Thermostat
Air Flow Unbalanced									●		●						Readjust Air Volume Dampers
System Undersized								●		●							Refigure Cooling Load
Broken Internal Parts												●					Replace Compressor
Broken Values												●					Test compressor Efficiency
Inefficient Compressor								●						●	●		Test Compressor Efficiency
High Pressure Control Open			●														Reset And Test Control
Unbalanced Pow er, 3PH		●				●	●										Test Voltage
Wrong Type Expansion Valve						●	●			●							Replace Valve
Expansion Valve Restricted						●	●	●		●			●	●			Replace Valve
Oversized Expansion Valve												●			●		Replace Valve
Undersized Expansion Valve						●	●	●		●			●				Repalce Valve
Expansion Valve Bulb Loose												●			●		Tighten Bulb Bracket
Inoperative Expansion Valve						●		●					●				Check Valve Operation
Loose Hold-dow n Bolts												●					Tighten Bolts

◆ Heat Pump Mode Only

(Courtesy Amana Corporation)

HEAT PUMP TROUBLESHOOTING GUIDE

COMPRESSOR DIAGNOSTICS

COMPRESSOR & OUTDOOR FAN MOTOR DO NOT START

1. NO POWER TO OUTDOOR UNIT
   A. Blown Fuse
   B. Tripped Circuit Breaker
   C. High Voltage Wiring
2. TRANSFORMER
   A. Improper Primary Voltage
   B. Improper Secondary Voltage
3. OPEN CONTROL CIRCUIT
   A. Blown 24V. Control Circuit Fuse
   B. Thermostat
   C. Low Voltage Wiring
4. DEFECTIVE CONTACTOR
   A. Coil Open, Shorted, Grounded
   B. Contacts Dirty or Burned
   C. Binding Mechanically
5. EQUIPMENT IMPROPERLY WIRED (FACTORY OR FIELD WIRING)

COMPRESSOR TRIES BUT FAILS TO START — TRIPS INTERNAL OVERLOAD (IOL)

1. OPEN START WINDINGS
2. SHORTED RUN WINDING
3. LOCKED ROTOR
4. LOW LINE VOLTAGE
5. DEFECTIVE CAPACITORS
   A. Run
   B. Start*
6. SYSTEM PRESSURES NOT EQUALIZED (ALLOW 5 MINUTES)
7. DEFECTIVE START RELAY (CSR)*
8. COMPRESSOR IMPROPERLY WIRED
   A. Loose Terminal Connection
   B. Open Wire
9. DEFECTIVE INTERNAL OVERLOAD (IOL)

*If Either of the Starting Components are Defective Both Should Be Changed.

(Courtesy The Trane Company)

HEAT PUMP TROUBLESHOOTING GUIDE

UNIT WILL NOT START AUTOMATICALLY, BUT WILL WHEN CONTACTOR (MS) IS MANUALLY CLOSED.

(USE INSULATED TOOL TO AVOID SHOCK HAZARD)

1. OPEN CONTROL CIRCUIT
   A. Blown 24V. Control Circuit Fuse
   B. Thermostat
   C. Low Voltage Wiring
2. DEFECTIVE TRANSFORMER
   A. Improper Primary Voltage
   B. Improper Secondary Voltage
3. DEFECTIVE CONTACTOR (MS)
   A. Coil Open, Shorted, Grounded
   B. Burned or Dirt Contacts
   C. Binding Mechanically

COMPRESSOR WON'T START, BUT OUTDOOR MOTOR RUNS.

1. OPEN INTERNAL OVERLOAD (IOL) (REFER TO "COMPRESSOR CYCLES ON IOL" PAGE 9)
2. BLOWN START WINDING FUSE
3. LOW LINE VOLTAGE
4. DEFECTIVE CAPACITOR
   A. Run
   B. Start*
5. START RELAY CONTACTS OPEN*
6. COMPRESSOR IMPROPERLY WIRED
   A. Loose terminal connection
   B. Open Wire
7. COMPRESSOR WINDINGS OPEN**

* If Either of the Starting Components are Defective Both Should Be Changed.

**Confirm by Checking Resistance from Start Terminal to Run Terminal.

(Courtesy The Trane Company)

HEAT PUMP TROUBLESHOOTING GUIDE

COMPRESSOR CYCLES ON INTERNAL OVERLOAD (IOL) (AFTER STARTING)

1. HIGH HEAD AND HIGH SUCTION PRESSURE
   - A. Dirty Condenser Coils
   - B. Overcharge
   - C. Restricted Air to Condenser
   - D. Filter Dirty (Heating Mode)
   - E. Lack of ID Airflow (Heating Mode)
   - F. Defective Condenser Fan Motor/Capacitor
     1. OD For Cooling
     2. ID For Heating
   - G. Recirculation of Condenser Air
   - H. Excessive Airflow through Evaporator (Cooling Mode)
   - I. High Return Air Registers
   - J. Heating Operation Above 65°F. OD Ambient.
2. DEFECTIVE RUN/START CAPACITORS
3. IMPROPER OR DEFECTIVE START RELAY (CSR)
4. **HIGH LINE VOLTAGE (TOLERANCE IS VOLTAGE RATING PLUS OR MINUS 10%)**
5. HIGH SUPERHEAT
   - A. Low Refrigerant Charge
   - B. Liquid Side Restriction — Drier, TXV, Quick Attach Fittings
   - C. Leaking Switch-Over Valve
   - D. Leaking Internal Pressure Relief Valve
   - E. Low Side Restriction — Metering Device, Distributor Tubes, Coils, Quick-Attach Fittings
   - F. Excessive Airflow through Evaporator (Cooling Mode)
   - G. High Superheat During Defrost
6. **TIGHT BEARINGS (IOL. WILL TRIP AFTER COMPR. STARTS)**
7. **LOW LINE VOLTAGE (TOLERANCE IS VOLTAGE RATING PLUS OR MINUS 10%)**
8. **OPEN START WINDINGS**
9. **SHORTED WINDINGS**
10. **DEFECTIVE INTERNAL OVERLOAD (IOL)**

(Courtesy The Trane Company)

HEAT PUMP TROUBLESHOOTING GUIDE

UNIT SHORT CYCLES

1. ROOM THERMOSTAT MOUNTED IN AIR STREAM
   - A. Location (Effected By Supply Air)
2. CONTROL CIRCUIT OPENS INTERMITTENTLY
3. HOLE BEHIND THERMOSTAT NOT PLUGGED
4. DEFECTIVE INTERNAL OVERLOAD (IOL)

COMPRESSOR RUNS CONTINUOUSLY — NO COOLING

1. ID FAN MOTOR INOPERATIVE
   - A. Fan Capacitor
   - B. Fan Relay Defective
   - C. Fan Motor Defective
2. INDOOR COIL ICED
   - A. Dirty Filter/Evaporator
   - B. Low Airflow
   - C. Low Refrigerant Charge
3. SWITCH-OVER VALVE (SOV)
   - A. Defective Coil
   - B. Stuck Mechanically
4. INADEQUATE PUMPING (LOW HEAD & HIGH SUCTION PRESSURE)

COMPRESSOR RUNS CONTINUOUSLY — INADEQUATE COOLING

1. IMPROPER EVAPORATOR AIRFLOW (LOW OR HIGH)
   - A. Restricted Duct Work
   - B. Dirty Filters/Coils
   - C. Registers And Grills Shut Off
   - D. Defective Motor/Capacitors
2. INADEQUATE CONDENSER AIRFLOW
   - A. Dirty Condenser
   - B. Improper Fan Rotation
   - C. Defective Motor/Capacitor
3. REFRIGERANT CHARGE
   - A. Low
   - B. Overcharge
4. LOW SIDE REFRIGERANT RESTRICTION
   - A. Metering Device
   - B. Distributor Tubes
   - C. Coils
   - D. Quick Attach Fittings
   - E. Brazed Connection
   - F. Suction Line
5. LIQUID LINE RESTRICTION
   - A. Drier
   - B. Metering Device
   - C. Quick Attach Fittings
   - D. Brazed Connections
6. INEFFICIENT COMPRESSOR (LOW HEAD & HIGH SUCTION)
7. LEAKING CHECK VALVE (INDOOR)

(Courtesy The Trane Company)

HEAT PUMP TROUBLESHOOTING GUIDE

8. LEAKING SWITCH-OVER VALVE (SOV)
9. COMPRESSOR CYCLING ON INTERNAL OVERLOAD (IOL) (SEE PAGE 9)
10. SUPPLEMENTARY HEATER ENERGIZED
    - A. Malfunctioning Heater Component
    - B. Malfunctioning Defrost Control (DFC)
11. EQUIPMENT
    - A. Undersized
    - B. Improperly Matched Indoor & Outdoor Sections
12. NON-CONDENSABLES IN SYSTEM
13. UNCONDITIONED AIR ENTERING RETURN AIR DUCT
14. POOR DUCT INSULATION
15. SUPPLY REGISTERS
    - A. Location
    - B. Type
    - C. Too Many
    - D. Too Few
16. POORLY DESIGNED DUCT SYSTEM

COMPRESSOR RUNS CONTINUOUSLY — NO HEATING

(THERMOSTAT NOT CALLING FOR RESISTANCE HEATERS)

1. SWITCH-OVER VALVE (SOV) STUCK MECHANICALLY
2. OUTDOOR COIL ICED UP
    - A. Dirty Evaporator
    - B. Low Refrigerant Charge
    - C. Excessive Water Dripping From Eaves
    - D. Defective Motor/Capacitor
    - E. Failure to Defrost Properly
3. INADEQUATE PUMPING (LOW HEAD & HIGH SUCTION)

COMPRESSOR RUNS CONTINUOUSLY — INADEQUATE HEATING

A. THERMOSTAT NOT CALLING FOR RESISTANCE HEATERS

1. IMPROPER CONDENSER AIRFLOW (LOW OR HIGH)
    - A. Restricted Ductwork
    - B. Dirty Filters
    - C. Registers and Grills Shut Off
    - D. Defective Motor/Capacitor
2. INADEQUATE EVAPORATOR AIRFLOW
    - A. Dirty Evaporator
    - B. Improper Fan Rotation
    - C. Defective Motor/Capacitor
    - D. Improper Defrosting
3. REFRIGERANT CHARGE
    - A. Low
    - B. Overcharge

(Courtesy The Trane Company)

HEAT PUMP TROUBLESHOOTING GUIDE

4. LOW SIDE REFRIGERANT RESTRICTIONS
   - A. Metering Device
   - B. Distributor Tubes
   - C. Coils
   - D. Quick Attach Fittings
   - E. Brazed Connections
5. LIQUID LINE RESTRICTION
   - A. Drier
   - B. Metering Device
   - C. Quick Attach Fittings
   - D. Brazed Connections
6. INEFFICIENT COMPRESSOR (LOW HEAD & HIGH SUCTION)
7. LEAKING CHECK VALVE (OUTDOOR)
8. LEAKING SWITCH-OVER VALVE (SOV)
9. COMPRESSOR CYCLING ON IOL (SEE PAGE 9)
10 EQUIPMENT
   - A. Undersized
   - B. Improperly Matched Indoor and Outdoor Sections.
11. UNCONDITIONED AIR ENTERING RETURN AIR DUCT
12. POOR DUCT INSULATION
13. NON-CONDENSABLES IN SYSTEM
14. SUPPLY REGISTERS
   - A. Location
   - B. Type
   - C. Too Many
   - D. Too Few
15. POORLY DESIGNED DUCT SYSTEM

COMPRESSOR RUNS CONTINUOUSLY — INADEQUATE HEATING

B. THERMOSTAT CALLING FOR RESISTANCE HEATERS

1. IMPROPER LOW VOLTAGE WIRING
2. OUTDOOR THERMOSTAT (ODT) CONTACTS NOT CLOSING
   - A. Improper Setting
   - B. Defective Outdoor Thermostat (ODT)
3. THERMOSTAT DEFECTIVE
4. SUPPLEMENTARY HEATER INOPERATIVE
   - A. Contactor/Sequencer Defective
   - B. Thermal Cut-Out Open
   - C. Fuse Links Open
   - D. Heater Element Open
5. SWITCH-OVER VALVE (SOV) STUCK IN COOLING POSITION
6. DEFROST CONTROL (DFC) WON'T TERMINATE
   - A. Timer Motor Inoperative
   - B. Control Mechanism Stuck

(Courtesy The Trane Company)

HEAT PUMP TROUBLESHOOTING GUIDE

OUTDOOR UNIT NOISY

1. LACK OF MOUNTING PADS UNDER EACH CORNER
2. UNIT NOT LEVEL
3. RATTLING SHEET METAL
4. ICE BUILT UP UNDER BASEPAN
5. COMPRESSOR NOISE
   A. Broken Springs
   B. Compressor Hitting Shell
   C. Discharge Line Hitting Shell
   D. Refrigerant Overcharge
   E. Compressor Mounting Bolts (Too Tight or Too Loose)

DEFROST SYSTEM DIAGNOSTICS

GE MORRISON DEFROST CONTROL PROBLEMS

EXCESSIVE ICE BUILD-UP ON OD COIL

1. LOW REFRIGERANT CHARGE
2. DEFROST CONTROL WON'T INITIATE
   A. No Voltage To Timer
   B. Control Timer Inoperative
   C. Control Contacts Not Closing To Defrost Relay
   D. Time Setting Is Between Points
   E. Poor Or No Contact Between Defrost Terminator And Lower Pass Of O.D. Coil
3. DEFROST RELAY INOPERATIVE
   A. Open Coil
   B. Contacts Not Making
   C. Inoperative Wiring
4. SOV INOPERATIVE
   A. Stuck In Heating Mode
   B. Open Switch-Over Valve (SOV) Coil
5. DEFROST TIME SETTING INADEQUATE FOR THE GEOGRAPHIC AREA
6. DEFROST CONTROL CONTACT TO OD FAN FAILS TO OPEN
7. DEFROST CONTROL TERMINATES, BUT DOES NOT REMOVE ICE
   A. Windy Conditions
   B. Outdoor Unit Located Under Eaves
   C. Lack Of Proper Drainage
   D. Night setback operation below 60°F.

(Courtesy The Trane Company)

HEAT PUMP TROUBLESHOOTING GUIDE

ICE BUILD UP ON LOWER PART OF OUTDOOR COIL

1. LOW REFRIGERANT CHARGE
2. DEFROST TERMINATOR
   A. Connected To Wrong Pass Of Outdoor Coil
   B. Poor Contact
3. LEAKING CHECK VALVE (OUTDOOR)
4. DISTRIBUTOR TUBE RESTRICTED
5. ONE PASS OF OD COIL RESTRICTED
6. LACK OF PROPER DRAINAGE

DEFROST INITIATES, BUT TERMINATES ONLY ON 10 MINUTE OVERRIDE

1. LOW REFRIGERANT CHARGE
2. DEFROST TERMINATOR CONTACTS STUCK CLOSED
3. WINDY CONDITIONS
4. NIGHT SETBACK
5. UNIT LOCATION
6. DEFROST TERMINATOR IN CONTACT WITH ICE

DEFROST CYCLE INITIATES, BUT WILL NOT TERMINATE

1. DEFROST RELAY STUCK
2. SWITCH-OVER VALVE STUCK IN COOLING MODE

UNIT GOES INTO DEFROST IN COOLING MODE

1. DEFROST TERMINATOR STUCK CLOSED

CONTROL TERMINATES DEFROST BEFORE FROST IS GONE

1. DEFROST TERMINATOR IS LOCATED ON WRONG PASS
2. REFRIGERANT OVERCHARGE

(Courtesy The Trane Company)

HEAT PUMP TROUBLESHOOTING GUIDE

RANCO DEFROST CONTROL PROBLEMS EXCESSIVE ICE BUILD-UP ON OD COIL

1. LOW REFRIGERANT CHARGE
2. DEFROST CONTROL WON'T INITIATE
   A. No Voltage To Timer
   B. Control Timer Inoperative
   C. Control Contacts Not Closing To Defrost Relay
   D. Time Setting Is Between Set Points
   E. Poor Bulb Contact (Milder Ambients Only)
3. DEFROST RELAY INOPERATIVE
   A. Open Coil
   B. Contacts Not Making
   C. Incorrect Wiring
4. SOV INOPERATIVE
   A. Stuck In Heating Mode
   B. Open Switch-Over Valve (SOV) Coil
5. DEFROST TIME SETTING INADEQUATE
6. DEFROST CONTROL CONTACT TO OD FAN FAILS TO OPEN
7. DEFROST CONTROL TERMINATES, BUT DOES NOT REMOVE ICE
   A. Windy Conditions
   B. OD Unit Located Under Eaves
   C. Lack Of Proper Drainage
   D. Night setback operation

ICE BUILT UP ON LOWER PART OF OD COIL

1. LOW REFRIGERANT CHARGE
2. DEFROST CONTROL (DFC) BULB
   A. Connected To Wrong Pass
   B. Poor Bulb Contact
3. LEAKING CHECK VALVE (OUTDOOR)
4. DISTRIBUTOR TUBE RESTRICTED
5. ONE PASS OF OD COIL RESTRICTED
6. LACK OF PROPER DRAINAGE

DEFROST INITIATES, BUT TERMINATES ONLY ON 10 MINUTE TIME OVERRIDE

1. LOW REFRIGERANT CHARGE
2. DEFROST CONTROL (DFC) BULB
   A. Loss Of Charge
   B. Poor Contact
   C. Connected To Wrong Pass
3. WINDY CONDITIONS
4. NIGHT SETBACK
5. UNIT LOCATION
6. DEFROST CONTROL (DFC) BULB CAPILL-ARY IN CONTACT WITH ICE
   A. Ice In Protective Sheath
   B. Routing Of Capillary Tube Through Icy Areas

(Courtesy The Trane Company)

HEAT PUMP TROUBLESHOOTING GUIDE

DEFROST CYCLE INITIATES, BUT WILL NOT TERMINATE

1. DEFROST RELAY STUCK
2. SWITCH-OVER VALVE STUCK IN COOLING MODE

UNIT GOES INTO DEFROST DURING COOLING MODE

1. DEFECTIVE DEFROST CONTROL (DFC)

CONTROL TERMINATES DEFROST BEFORE FROST IS GONE

1. DEFROST CONTROL BULB ON WRONG PASS
2. REFRIGERANT OVERCHARGE

ROBERTSHAW DEFROST CONTROL PROBLEMS

EXCESSIVE ICE BUILD-UP ON OUTDOOR COIL

1. RUBBER HOSE TO DRIP LEG LOOSE
2. ATMOSPHERIC TUBE NOT CONNECTED TO DEFROST CONTROL
3. RUPTURED DRIP LEG
4. VACUUM TUBE BELL NOT FACING DOWN CLOSE TO THE SLINGER
5. SENSING SWITCH ADJUSTMENT SET TOO HIGH
6. SENSING TUBE BLOCKAGE
   A. Vacuum Sensing Tube
   B. Atmospheric Sensing Tube (Positive Pressure Reference)
7. DEFROST RELAY DEFECTIVE
   A. Coil Open
   B. Contacts Defective
   C. Wired Wrong
8. SWITCH-OVER VALVE (SOV) STUCK MECHANICALLY IN HEATING MODE
9. BULB OUT OF WELL (MILDER AMBIENTS ONLY)
10. TERMINATION BULB CRUSHED, DENTED, BENT, CAUSING WRONG BULB CALIBRATION
11. TOO MUCH AIR IS BYPASSING O.D. COIL (EITHER THROUGH COIL OR UNDER/OVER COIL)
12. DEFECTIVE DEFROST CONTROL (DFC)
13. DEFROST CONTROL (DFC) WIRED WRONG
14. DEFECTIVE OD FAN MOTOR/ CAPACITOR

(Courtesy The Trane Company)

HEAT PUMP TROUBLESHOOTING GUIDE

ICE BUILD UP ON LOWER PART OF OD COIL

1. LOW REFRIGERANT CHARGE
2. POOR DEFROST CONTROL (DFC) BULB CONTACT
3. LEAKING CHECK VALVE (OUTDOOR)
4. DISTRIBUTOR TUBE RESTRICTED
5. ONE PASS OF OD COIL RESTRICTED
6. LACK OF PROPER DRAINAGE

DEFROST CYCLE INITIATES, BUT WILL NOT TERMINATE

1. LOW GAS CHARGE
2. LOST CHARGE IN TERMINATION BULB
3. SWITCH-OVER VALVE (SOV) STUCK IN COOLING MODE
4. TEMPERATURE SETTING TOO HIGH
5. WINDY CONDITIONS
6. BULB OUT OF WELL
7. NIGHT SETBACK OPERATION

DEFROST CYCLE INITIATES — NO ICE ON COIL

1. CHARGE LOST IN TERMINATION BULB (DFC)
2. DEFROST CONTROL WIRED WRONG
3. DIRTY OUTDOOR COIL
4. PRESSURE SETTING SET TOO LOW

UNIT GOES INTO DEFROST DURING COOLING MODE

1. DEFECTIVE DEFROST CONTROL (DFC)

DEFROST CYCLE OCCURS TOO OFTEN

1. DEFROST CONTROL PRESSURE SETTING SET TOO LOW
2. DIRTY OUTDOOR COIL
3. WINDY CONDITIONS

CONTROL TERMINATES DEFROST BEFORE FROST IS GONE

1. RUBBER SLEEVE MISSING FROM DEFROST CONTROL BULB
2. REFRIGERANT OVERCHARGE

(Courtesy The Trane Company)

HEAT PUMP TROUBLESHOOTING GUIDE

DWYER DEFROST CONTROL PROBLEMS

EXCESSIVE ICE BUILD-UP ON OUTDOOR COIL

1. DEFROST TERMINATION (DT) SWITCH
   A. Not Closed
   B. Poor Contact
2. SENSING TUBES
   A. Vacuum Sensing Tube Blocked Or Broken
   B. Atmospheric Sensing Tube Blocked Or Broken
3. VACUUM TUBE BELL NOT SENSING PROPERLY (MOUNTING)
4. DIAPHRAM SWITCH (DS) CONTACTS DO NOT CLOSE
5. DEFROST RELAYS DEFECTIVE (D & DR)
   A. Coils Open
   B. Contacts Dirty Or Burned
   C. Wired Wrong
6. SWITCH-OVER VALVE STUCK MECHANICALLY IN HEATING MODE
7. DIAPHRAM SWITCH
   A. Calibration
   B. Defective Control
8. OUTDOOR FAN MOTOR(S)
   A. Rotation
   B. Defective Motor/Capacitor
   C. Dual Fans (Both Must Be Operating)

DEFROST CYCLE INITIATES, BUT WILL NOT TERMINATE

1. LOW REFRIGERANT CHARGE
2. DEFROST TERMINATION (DT) CONTROL DOES NOT OPEN
3. TERMINATION CONTROL NOT PROPERLY MOUNTED ON MOUNTING PLATE
4. SWITCH-OVER VALVE (SOV) STUCK MECHANICALLY IN COOLING MODE
5. DEFROST RELAYS (D,DR) STUCK CLOSED
6. WINDY CONDITIONS
7. NIGHT SETBACK OPERATION

DEFROST CYCLE INITIATES — NO ICE ON COIL

1. DIAPHRAM SWITCH (DS) CONTACTS STUCK CLOSED
2. DIRTY OUTDOOR COIL
3. WINDY CONDITIONS

DEFROST OCCURS TOO OFTEN

1. DIAPHRAM SWITCH (DS) CONTACTS STUCK
2. DIAPHRAM CONTROL PRESSURE SETTING SET TOO LOW
3. WINDY CONDITIONS
4. DIRTY OUTDOOR COIL

DEFROST TERMINATES BEFORE FROST IS GONE

1. DEFROST TERMINATION (DT) SWITCH DEFECTIVE
2. REFRIGERANT OVERCHARGE

(Courtesy The Trane Company)

HEAT PUMP TROUBLESHOOTING GUIDE

DEMAND DEFROST CONTROL PROBLEMS

EXCESSIVE ICE BUILT-UP ON OD COIL

1. LOW REFRIGERANT CHARGE
2. DEFROST CONTROL WILL NOT INITIATE
   - A. No 24 VAC Between R&B At Defrost Control
   - B. No 24 VAC Between B&Y At Defrost Control With System Running
   - C. Verify Correct Sensor Location, Mounting And Their Resistance
   - D. Verify Ambient Sensor Is Connected To AMB Position On Defrost Control
   - E. Verify Coil Sensor Is Connected To Coil Position On Defrost Control
3. SOV INOPERATIVE
   - A. Stuck In Heating Mode
   - B. Open Switchover Valve (SOV) Coil
   - C. Defective Defrost Control
4. DEFROST CONTROL CONTACTS TO OD FAN FAIL TO OPEN DURING DEFROST CYCLE
5. DEFROST CONTROL TERMINATES, BUT DOES NOT REMOVE ICE
   - A. Windy Conditions
   - B. Outdoor Unit Located Under Eaves
   - C. Lack Of Proper Drainage
   - D. Night Setback Operation

ICE BUILD UP ON LOWER PART OF OUTDOOR COIL

1. LOW REFRIGERANT CHARGE
2. COIL SENSOR CONNECTED TO WRONG PASS OF OUTDOOR COIL, OR POOR CONTACT
3. LEAKING CHECK VALVE (OUTDOOR UNIT)
4. DISTRIBUTOR TUBE RESTRICTED
5. ONE PASS OF OD COIL RESTRICTED
6. LACK OF PROPER DRAINAGE

DEFROST INITIATES, BUT TERMINATES ONLY ON A TIME OVERRIDE

1. LOW REFRIGERANT CHARGE
2. OUTDOOR FAN ON DURING DEFROST
3. WINDY CONDITIONS
4. NIGHT SETBACK OPERATION
5. UNIT LOCATION
6. COIL SENSOR IN CONTACT WITH ICE
7. COIL SENSOR CIRCUIT OPEN OR READING VERY HIGH RESISTANCE.

DEFROST CYCLE INITIATES, BUT WILL NOT TERMINATE

1. DEFROST CONTROL
2. SWITCH-OVER VALVE STUCK IN COOLING MODE

UNIT GOES INTO DEFROST IN COOLING MODE

1. DEFECTIVE SENSORS
2. DEFECTIVE DEFROST CONTROL

(Courtesy The Trane Company)

HEAT PUMP TROUBLESHOOTING GUIDE

CONTROL TERMINATES DEFROST BEFORE FROST IS GONE

1. COIL SENSOR MOUNTED IN WRONG LOCATION OR HAS INCORRECT RESISTANCE READING
2. REFRIGERANT OVERCHARGE

DEFROST INITIATES ABOUT EVERY 15 MINUTES

1. COIL SENSOR
2. AMBIENT SENSOR
3. DEFROST CONTROL

DEFROST INITIATES ABOUT EVERY 30 MINUTES. FAULT LIGHT ON INDOOR THERMOSTAT WILL BE FLASHING IF WIRED

1. COIL SENSOR
2. AMBIENT SENSOR
3. WEATHER CONDITIONS
4. NIGHT SETBACK OPERATION
5. OUTDOOR FAN ON DURING DEFROST
6. SYSTEM REFRIGERANT CHARGE
7. SOV OPERATION

REFRIGERANT SYSTEM DIAGNOSTICS

IN HEATING MODE THE OD COIL MUST BE CLEAR OF ICE

IMMEDIATE TRIPPING OF INTERNAL PRESSURE RELIEF VALVE (IPR) ON START UP

1. DISCHARGE SIDE RESTRICTION
   A. Discharge Tube
   B. Muffler
   C. Switch-Over Valve (SOV) Not Shifting Properly
   D. Copper/Aluminum Joint Blocked (Condenser Inlet)

HEAD PRESSURE HIGHER THAN PERFORMANCE CHART

1. OUTDOOR AIR SYSTEM (COOLING)
   A. Dirty Outdoor Coil
   B. Restricted Inlet Air
   C. Recirculation Of Condenser Air
   D. Defective Motor/Capacitor
2. INDOOR AIR SYSTEM (HEATING)
   A. Dirty Filters
   B. Dirty Blower
   C. Dirty Indoor Coil
   D. Supply Registers Closed
   E. Inadequate Supply Registers
   F. Inadequate Return Air Grill Area
   G. Restricted Ductwork
   H. Defective Motor/Capacitor

(Courtesy The Trane Company)

HEAT PUMP TROUBLESHOOTING GUIDE

3. OVERCHARGE OF REFRIGERANT
4. PARTIAL RESTRICTIONS
   A. Drier
   B. Condenser Coils
   C. Connecting Tubing (Heating)
   D. Quick Attach Fittings
5. EXCESSIVE EVAPORATOR LOAD
   A. Dry Bulb Temp
   B. Wet Bulb Temp
6. HIGH INLET AIR TEMP. TO CONDENSER (HEATING)
7. NON-CONDENSABLES IN SYSTEM
8. WRONG SIZE ACCUTRON FLOW CONTROL INSTALLED

HEAD PRESSURE LOWER THAN PERFORMANCE CHART

1. UNDER CHARGE OF REFRIGERANT
2. LOW INLET AIR TEMP. TO CONDENSER (HEATING)
3. OUTDOOR AIR SYSTEM (HEATING)
   A. Dirty Outdoor Coil
   B. Restricted Inlet Air
   C. Recirculation Of Outdoor Air
   D. Defective Motor/Capacitor
4. TOTAL/PARTIAL RESTRICTION
   A. Drier
   B. Metering Device
   C. Coils
   D. Liquid Line
   E. Quick Attach Fittings
   F. Suction Line (Cooling Mode)
5. LEAKING CHECK VALVE
6. EXCESSIVE CONDENSER AIRFLOW (HEATING)
7. THERMOSTATIC EXPANSION VALVE (TXV) STUCK OPEN
8. SWITCH-OVER VALVE (SOV) LEAKING
9. INEFFICIENT COMPRESSOR (LOW HEAD & HIGH SECTION)
10. INTERNAL PRESSURE RELIEF VALVE
11. INDOOR AIR SYSTEM (COOLING)
    A. Dirty Filters
    B. Dirty Blower
    C. Dirty Indoor Coil
    D. Supply Registers Closed
    E. Inadequate Supply Registers
    F. Inadequate Return Air Grill Area
    G. Restricted Ductwork
    H. Defective Motor/Capacitor
    I. Low Evaporator Inlet Temp.

SUCTION PRESSURE HIGHER THAN PERFORMANCE CHART

1. OUTDOOR AIR SYSTEM (COOLING)
   A. Dirty Condenser Coil
   B. Restricted Inlet Air
   C. Recirculation Of condenser Air
   D. Defective Motor/Capacitor
2. OVERCHARGE OF REFRIGERANT
3. EXCESSIVE EVAPORATOR LOAD
   A. Dry Bulb Temp.
   B. Wet Bulb Temp.
4. SWITCH-OVER VALVE (SOV) LEAKING

(Courtesy The Trane Company)

HEAT PUMP TROUBLESHOOTING GUIDE

5. LEAKING CHECK VALVE (COOLING)
6. THERMOSTATIC EXPANSION VALVE (TXV) STUCK OPEN (COOLING)
7. INEFFICIENT COMPRESSOR (LOW HEAD & HIGH SUCTION)
8. INDOOR AIR SYSTEM (HEATING)
   A. Dirty Filters
   B. Dirty Blowers
   C. Dirty Coil
   D. Supply Registers Closed
   E. Inadequate Supply Registers
   F. Inadequate Return Air Grill Area
   G. Restricted Ductwork
   H. Defective Motor/Capacitor
9. NOT A MATCHED SYSTEM OR WRONG SIZE METERING DEVICE (COOLING)

SUCTION PRESSURE <u>LOWER</u> THAN PERFORMANCE CHART

1. LOW REFRIGERANT CHARGE
2. INDOOR AIR SYSTEM (COOLING)
   A. Dirty Filters
   B. Dirty Blower
   C. Dirty Indoor Coil
   D. Supply Registers Closed
   E. Inadequate Supply Registers
   F. Inadequate Return Air Grill Area
   G. Restricted Ductwork
   H. Defective Motor/Capacitor
   I. Low Evaporator Inlet Temp (Dry Bulb or Wet Bulb)
3. OUTDOOR AIR SYSTEM (HEATING)
   A. Dirty Outdoor Coil
   B. Restricted Inlet Air
   C. Recirculation Of Outdoor air
   D. Defective Motor/Capacitor
4. TOTAL/PARTIAL RESTRICTION
   A. Drier
   B. Metering Device
   C. Distributor Tubes
   D. Coils
   E. Connecting Tubes
   F. Quick Attach Fittings
5. LEAKING CHECK VALVE (HEATING)
6. THERMOSTATIC EXPANSION VALVE (TXV) STUCK OPEN (HEATING)
7. NON-CONDENSABLES IN SYSTEM
8. OUTDOOR UNIT OVERSIZED COMPARED TO INDOOR UNIT
9. EXCESSIVE CONDENSER AIRFLOW (HEATING)

GENERAL PROBLEM DIAGNOSTICS

EXCESSIVE OPERATING COSTS (HEATING OR COOLING)

1. LOW REFRIGERANT CHARGE
2. HEAT STRIPS ONLY OPERATING (HEAT PUMP CYCLING OFF)
3. HIGH HEAD PRESSURE
   A. Restricted Air To Condenser
   B. Overcharge
   C. Dirty Condenser Coils
   D. Dirty Filters

(Courtesy The Trane Company)

HEAT PUMP TROUBLESHOOTING GUIDE

E. Defective Condenser Fan Motor/Capacitor

4. I.D. CHECK VALVE LEAKING (COOLING)
5. O.D. CHECK VALVE LEAKING (HEATING)
6. DEFROST
   - A. Improper Operation
   - B. Occurs Too Frequently
7. HIGH AIRFLOW I.D.
8. EQUIPMENT SIZING
   - A. Undersized
   - B. Oversized
9. HEAT STRIPS OPERATING WITH COMPRESSOR IN MILD O.D. AMBIENTS
10. UNCONDITIONED AIR ENTERING RETURN AIR
11. EXCESSIVE DUCT LOSSES
12. THERMOSTAT LOCATION
13. NIGHT SETBACK OPERATION (NOT RECOMMENDED FOR HEAT PUMPS UNLESS DIGITAL THERMOSTAT IS USED)
14. HIGH CEILING RETURNS
15. NOT A MATCHED SYSTEM OR WRONG SIZE METERING DEVICE (COOLING)

COLD AIR COMPLAINTS (HEATING)

1. INDOOR AIRFLOW TOO HIGH
2. LOW REFRIGERANT CHARGE
3. SUPPLY REGISTERS
   - A. Location
   - B. Type
   - C. Too Many
   - D. Too Few
4. SUPPLEMENTARY HEAT DOES NOT COME ON DURING DEFROST.
5. SUPPLEMENTARY HEAT INOPERATIVE
   - A. Low Voltage Wiring
   - B. Defective Heater Components
   - C. Defective Outdoor Thermostat (ODT)
6. SWITCH-OVER VALVE (SOV) STUCK IN DEFROST
7. EXCESSIVE DUCT LOSSES
8. BREAK IN RETURN OR SUPPLY DUCT SYSTEM
9. POORLY DESIGNED DUCT SYSTEM
10. HEAT PUMP UNDERSIZED

SWITCH-OVER VALVE DOES NOT SWITCH

1. COMPRESSOR NOT OPERATIVE
2. INADEQUATE VOLTAGE TO COIL
3. SWITCH-OVER VALVE COIL DEFECTIVE
4. DEFROST (D) RELAY INOPERATIVE
   - A. Open Coil
   - B. Dirty Or Burned Contacts
   - C. Wired Wrong
   - D. Terminals, Connectors Loose Or Broken

(Courtesy The Trane Company)

ELECTRONIC AIR CLEANER TROUBLESHOOTING CHART

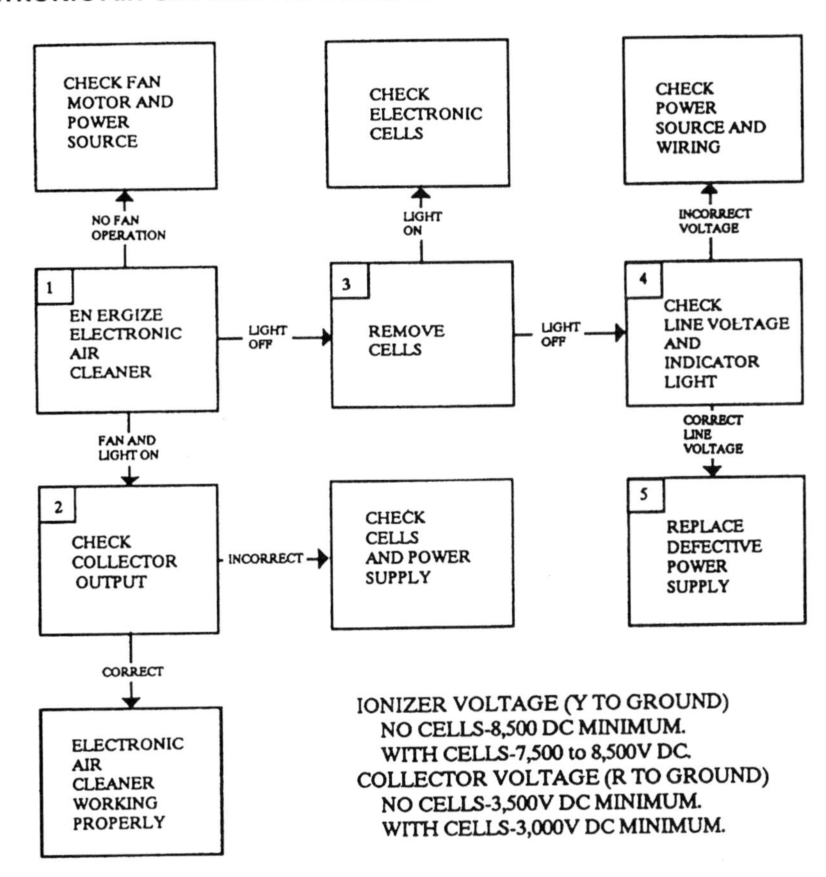

(Courtesy Smokemaster)

INDUSTRIAL ELECTRONIC AIR CLEANER TROUBLESHOOTING CHART

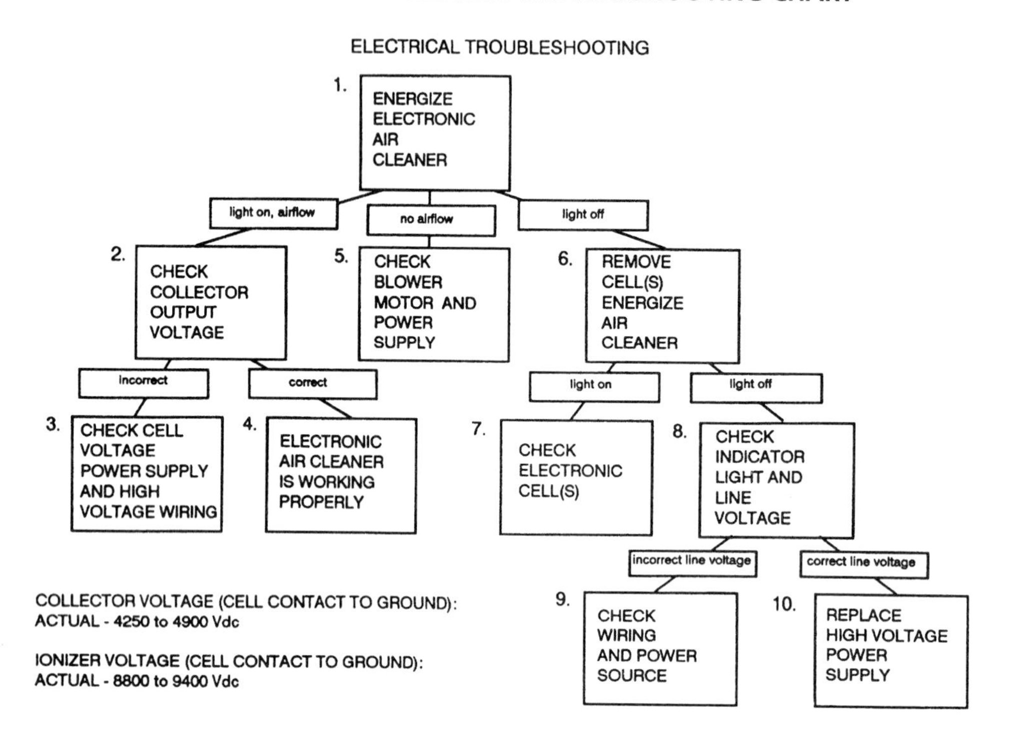

(Courtesy Smokemaster)

INDEX

P

R

S